Paths to a Culture of
Tolerance and Peace

RIVER PUBLISHERS SERIES IN CHEMICAL, ENVIRONMENTAL, AND ENERGY ENGINEERING

Series Editors

MEDANI P. BHANDARI
Akamai University, USA; Sumy State University, Ukraine and Atlantic State Legal Foundation, NY, USA

JACEK BINDA
PhD, Vice Rector of the International Affairs, Bielsko-Biala School of Finance and Law, Poland

DURGA D. POUDEL
PhD, University of Louisiana at Lafayette, Louisiana, USA

SCOTT GARNER
JD, MTax, MBA, CPA, Asia Environmental Holdings Group (Asia ENV Group), Asia Environmental Daily, Beijing/Hong Kong, People's Republic of China

HANNA SHVINDINA
Sumy State University, Ukraine

ALIREZA BAZARGAN
NVCo and University of Tehran, Iran

Indexing: All books published in this series are submitted to the Web of Science Book Citation Index (BkCI), to SCOPUS, to CrossRef and to Google Scholar for evaluation and indexing.

The "River Publishers Series in Chemical, Environmental, and Energy Engineering" is a series of comprehensive academic and professional books which focus on Environmental and Energy Engineering subjects. The series focuses on topics ranging from theory to policy and technology to applications.

Books published in the series include research monographs, edited volumes, handbooks and textbooks. The books provide professionals, researchers, educators, and advanced students in the field with an invaluable insight into the latest research and developments.

Topics covered in the series include, but are by no means restricted to the following:

- Energy and Energy Policy
- Chemical Engineering
- Water Management
- Sustainable Development
- Climate Change Mitigation
- Environmental Engineering
- Environmental System Monitoring and Analysis
- Sustainability: Greening the World Economy

For a list of other books in this series, visit www.riverpublishers.com

Paths to a Culture of Tolerance and Peace

Editors

Basma EL Zein
University of Business and Technology, Saudi Arabia

and

Ahmed Al Jarwan
Global Council for Tolerance and Peace, Malta

LONDON AND NEW YORK

Published 2021 by River Publishers
River Publishers
Alsbjergvej 10, 9260 Gistrup, Denmark
www.riverpublishers.com

Distributed exclusively by Routledge
4 Park Square, Milton Park, Abingdon, Oxon OX14 4RN
605 Third Avenue, New York, NY 10158

First published in paperback 2024

Paths to a Culture of Tolerance and Peace / by Basma EL Zein, Ahmed Al Jarwan.

Routledge is an imprint of the Taylor & Francis Group, an informa business

Publisher's Note
The publisher has gone to great lengths to ensure the quality of this reprint but points out that some imperfections in the original copies may be apparent.

While every effort is made to provide dependable information, the publisher, authors, and editors cannot be held responsible for any errors or omissions.

ISBN: 978-87-7022-208-2 (hbk)
ISBN: 978-87-7004-318-2 (pbk)
ISBN: 978-1-003-33904-5 (ebk)

DOI: 10.1201/9781003339045

Contents

Preface

The word peace does not have a unique meaning in the history of humanity, and it is convenient to assume that the challenge of unifying it for our time is extremely urgent. To understand this situation, we should keep in mind that, in the Roman sense (*pax*), it referred to the submission of any region or people to the Empire - which submitted them to obedience to the "*jus gentium*", meaning that peace was translated into obedience and submission.

Westernization defined the colonialism of several powers but highlighted the British *Pax* that took on a vast expression in the Commonwealth. In an attempt to reduce , the meanings of the word "peace", it ended up becoming defined as, "absence of war", which did not always mean loyalty in relations between different powers, but it meant just not appealing to violence that seeks, through the submission of different peoples and lands, the imposition of victory.

It was this vision that guided what was called *Pax Dei*, an expression sometimes intended to protect only non-combatants in times of war, but the most distinguished evaluators of the maintenance of peace between the States, being these habitual interveners in the action, although they cannot forget internal wars of a State, for differences of principles, values, or interests, are several and always valuable. It is above all the analysts of the faith, and creditors of international law, or very articulate university professors — for example at the Universities of Coimbra and Evora in Portugal, and Salamanca in Spain, who at the time of westernization by navigation and conquests, reminded a general thought of the rights of all Peoples encountered, a teaching that today is called "Iberian Peace Policy", inscribed in the "Intangible Heritage of Humanity", entrusted to UNESCO. Among the religions, Saint Augustine (354–430) stands out, who, in the "City of God", maintains that true peace does not depend only on the lack of hostilities, but rather on "tranquillity in order".

However, the most globalized thought was that of Kant, who advocated for the submission of the federation of free States to an obeyed law. This vision led to Jefferson's Universal Declaration of Rights (1776) entitled

Virginia Declaration of Rights, and, at the end of the 1939- 1945 war, to the UN General Declaration of Rights.

However, being in this period the first time when globalism implies the meeting of all cultures and beliefs of humanity – not forgetting the fragility of obedience to all UN bodies – it seems clear to me that two principles are in force, not written and fundamental elements that ensure, if observed, peace and sustainable development: "the single world", that is, without wars, and the "common home of humanity", that is, all with equal dignity and peace, thus allowing for sustained equal development.

In the UN Charter, there is no guiding reference on the cultural encounter of the various religions. However, there was an intervention that grew in importance. The illustrious Secretary- General, Dag Hammarskjöld, who was to be a victim of an attack in the Belgian Congo, created a modest room at the UN headquarters, with rows of modest chairs near the walls and a stone altar receiving light from above, that was called the Meditation Room for all religions. This intention found no visible support, but the UN General Assembly welcomed Catholic Popes, Bishops of Rome, including Paul VI, John Paul II, Pope Emeritus, and presently Pope Francis.

Also the German Father, Hans Küng, with his remarkable work on Christianity, Islam and Judaism that led to the creation of his Foundation for Global Ethics, affirms that, in the "religious situation of the time, there is no peace between Nations or peace between religions, without dialogue between religions; and there is no dialogue between religions without researching the good between religions. The existence of the Global Council for Tolerance and Peace is a valuable contribution to making this often abandoned objective finally part of the cultural heritage of Humanity.

ADRIANO MOREIRA

President of the Institute for Higher Studies of the Academy of Sciences of Lisbon Professor Emeritus of the Technical University of Lisbon

List of Figures

List of Tables

List of Abbreviations

BCE	Before the Common Era (B.C.)
EU	European Union
Experiences IN	Inclusive, Informal and Inclusive
GCTP	Global council for Tolerance and Peace
GTI	The Global Terrorism Index
H variety	High (language) variety
L variety	Low (language) variety
LWCs	Languages of Wider Communication
NATO	The North Atlantic Treaty Organization
OSCE	Organization for Security and Co-operation in Europe
SIL	Summer Institute of Linguistics (SIL International)
SP-EEI	Starting Point - Educational Experiences IN
UAE	United Arab Emirates
UN	United Nations
VUCA	Volatil, uncertain, complex and ambiguous

1

Human Fraternity Document and the Role of the United Arab Emirates

H.E. Ahmed Al Jarwan

Global Council for Tolerance and Peace

1.1 Introduction

In a world where social injustice, corruption, inequality, moral decline, terrorism, discrimination, extremism, and violence have spread, it is time to become united to cherish peace and work for its establishment. It is time to take in new strategies and new methods at the global level. It is time to act.

There is still hope in a bright future for all human beings, if we have faith in God, if we unite, and if we work together to advance the culture of mutual respect, to encourage dialogue, to defend the protection of belief, to defend justice, and to protect, empower, and educate women, children, and disabled persons.

A new milestone in the history of mankind has been achieved on 4 February 2019, the day where the Human Fraternity document has been released and signed by His Holiness Pope Francis and The Grand Imam of Al-Azhar Ahmad Al-Tayyeb in Abu Dhabi, UAE, the land of Tolerance.

1.2 UAE History with Tolerance and Peace

Being the beacon of tolerance and peace, since its foundation by Sheikh Zayed bin Sultan Al Nahyan, UAE was chosen to host this historical event.

UAE has been founded on the concepts of tolerance, intellectual exchange, and cultural and economic openness. It promotes acceptance and understanding as its core values. UAE is considered as a symbol of peace and security, especially that it is located in a region shaken by wars, discrimination, and conflict.

Figure 1.1   Sheikh Zayed Bin Sultan Al Nahyan.

Figure 1.2   Sheikh Mohammad Bin Zayed Al Nahyan.

His Highness Sheikh Mohammad Bin Zayed Al Nahyan, Crown Prince of Abu Dhabi and Deputy Supreme Commander of the UAE Armed Forces once said that UAE is continuing its traditional approach to supporting the efforts and initiatives to help the people of the region and promote tolerance,

Figure 1.3 His Holiness Pope Francis and The Grand Imam of Al-Azhar Ahmad Al-Tayyeb in Abu Dhabi.

coexistence, brotherhood, and mutual respect around the world, as it believes in the importance of these values.

He also said that "Undoubtedly, high moral values help maintain the stability and balance of societies, because, no matter how developed nations become, their development and success will always be fragile if it is not based on a solid moral grounding. The UAE has always been a nurturing home for tolerance and a deep-rooted value system, which we work hard to instill in our current and future generations."

UAE is playing a pioneering role by being a role model in promoting tolerance and building peace through different projects, programs, activities, and initiatives such as the following:

- considering year 2019 as the year of tolerance;
- establishing the first ministry of tolerance in the world;
- providing international assistance to people with disasters;
- supporting projects in construction, education, health, and humanitarian assistance in many countries;
- embracing the values of tolerance, peace, security, and multi-culturalism with more than 200 nationalities living in mutual respect and enjoying a decent quality of life;
- conducting conferences;
- running sports tournaments;
- offering awards to recognize peace builders and tolerance advocators;
- setting the National Tolerance Program;

- establishing the Abrahamic Family House, after Abraham the revered prophet in Islam, Judaism, and Christianity, to symbolize the state of coexistence and human fraternity;
- establishing the first multi-faith prayer room at Abu Dhabi International Airport;
- boasting more than 40 churches of different Christian denominations, as well as Sikh and Hindu temples;
- setting up anti-discrimination and anti-hate laws and centers to counter extremism;
- following up on the progress of the activation of the Document of the Human Fraternity;
- extolling the virtue of tolerance as it comes from an intrinsic part of the Islamic culture;
- observing tolerance at all levels: individual, organizational, national, and institutional;
- and many others.

All of this makes the release and the signature of the Human Fraternity document, in attendance of more than 400 religious' leaders, a powerful message to all mankind for living together and for a world of peace.

1.3 Insights on the Human Fraternity Document

It is a message to all of us, inviting us to promote, evolve, act, cooperate, and endure toward a happier, stable, and peaceful world.

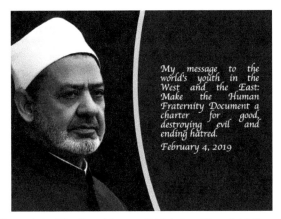

Figure 1.4   The Grand Imam of Al-Azhar Ahmad Al-Tayyeb.

The Grand Imam of Al-Azhar Ahmad Al-Tayyeb said on the occasion of the signing of the Human Fraternity document, "My message to the world's youth in the West and the East: Make the Human Fraternity Document a charter for good, destroying evil and ending hatred."

On the other hand, His Holiness Pope Francis also said, "The Document on Human Fraternity, which I signed today in Abu Dhabi with my brother The Grand Imam of Al-Azhar, invited all persons who have faith in God and faith in Human Fraternity to unite and work Together."

Taking in consideration the level of poverty, conflict, suffering of human beings, and distancing from our religious values — along with the feelings of frustration, isolation, desperation, injustice, lack of equitable distribution of natural resources, extremism, and many other issues that have been clearly mentioned in the Human Fraternity documents — we should not keep silent; we should unite and act to resolve all these issues, through dialogue, cooperation, and good education system, and the empowerment, protection, and support of women, children, elderly, disabled, and oppressed people.

All what has been achieved and what peace builders and tolerance advocators are working on nowadays are synchronized with many initiatives and programs proposed by many international entities such as UN, UNESCO, UNFPA, and many others.

The Document is considered a road map for global peace and coexistence and sets different principles, summarized in Figures 1.5 and 1.6.

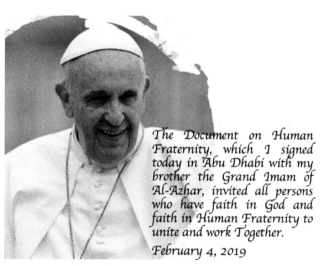

The Document on Human Fraternity, which I signed today in Abu Dhabi with my brother the Grand Imam of Al-Azhar, invited all persons who have faith in God and faith in Human Fraternity to unite and work Together.

February 4, 2019

Figure 1.5 His Holiness Pope Francis.

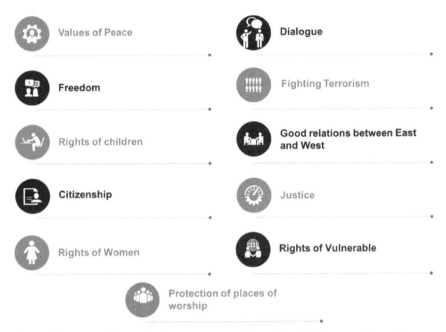

Figure 1.6 Human Fraternity document as a road map for global peace and coexistence.

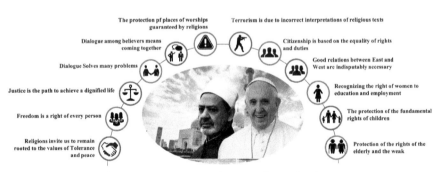

Figure 1.7 Key principles of Human Fraternity document.

All programs, initiatives, activities, meetings, and projects to resolve these issues should lead to a world of peace, justice, goodness, beauty, and coexistence.

It is the responsibility of everyone — politicians, leaders, intellectuals, scientists, technologists, educators, philosophers, religious figures, artists, media professionals, and women and men of different cultures — to work

Al-Azhar and the catholic church as that this Document become the object of Research and Reflection in all schools, universities and institutes of formation, thus helping to educate new generations to bring goodness and peace to others, and to be defenders everywhere of the rights of the oppressed and the least of our brothers and sisters.

Figure 1.8 The Human Fraternity document to become the object of research and reflection in all schools, universities, and institutes of formation.

together, united hand by hand to build the human values, to promote tolerance and peace everywhere, and to develop with sustainability to empower human beings; and this is what peace builders and tolerance advocators are doing. Everyone should work on actualizing the Human Fraternity document through our academic and social programs, our projects serving women and youth, sustainable development, society and community building, media and communication, our meetings that are open for dialogue, brainstorming, negotiation and discussion, our collaborations, and our conferences and symposiums to develop and revive the moral values and coexistence in human beings, and minimize these catastrophic crises.

From educational perspectives, His Holiness Pope Francis and The Grand Imam of Al-Azhar Ahmad Al-Tayyeb asked all schools, universities, and academic institutions to study the document, conduct applied research, and seek means of incorporating its principles and objectives in their curriculum to build a new generation of peace builders and tolerance advocators.

Figure 1.9   H. E. Antonio Guterres.

António Guterres, the United Nations General Secretary, commented on the Human Fraternity document saying that "the Human Fraternity Document, signed by The Grand Imam and Pope Francis, represents the practical application of religious tolerance and respect, and a direct and intended message to all believers that diversity in religion is a divine wisdom, just like difference in color, gender and language. I look forward to the realization of the efforts of the Committee through meetings with religious leaders and heads of international organizations and other figures, in addition to various initiatives that would spread peace and love among all human beings."

He also added that "the signing of this document by the world's top religious figures reflects the universality of its message that involves necessity of respecting and protecting the principle of religious freedom. This couldn't have been possible without the support and efforts of His Highness Sheikh Mohamed bin Zayed Al Nahyan, Crown Prince of Abu Dhabi and Deputy Supreme Commander of the UAE Armed Forces, in getting this document signed and implemented to promote global coexistence and understanding. I also admire the diversity within the Committee entrusted with achieving the document's objectives, as it contains representatives of different religions and nationalities."

1.4 Higher Committee of Human Fraternity

The Higher Committee of Human Fraternity comprises a diverse set of international religious leaders, educational scholars, and cultural leaders who were inspired by the Document on Human Fraternity and are dedicated to sharing its message of mutual understanding and peace.

The board includes those from the UAE, Spain, Italy, Egypt, and the USA and will expand to incorporate leaders of other denominations and beliefs in the coming years.

Members of the higher committee include:

1. Miguel Guixot, President of the Pontifical Council for Interreligious Dialogue of the Holy See, who has been appointed as the role of Chairman;
2. Judge Mohamed Abdel Salam, Advisor to The Grand Imam, who acts as Committee Secretary;
3. Professor Mohamed Mahrasawi, President of Al-Azhar University;
4. Monsignor Yoannis Gaid, Personal Secretary of Pope Francis;
5. Mohamed Al Mubarak, Chairman of the Department of Culture and Tourism — Abu Dhabi;
6. Dr. Sultan Al Rumaithi, Secretary-General of the Muslim Council of Elders;
7. Yasser Al Muhairi, an Emirati writer;
8. Rabbi Bruce Lustig, a Senior Rabbi at Washington Hebrew Congregation.

Figure 1.10   The Higher Committee of Human Fraternity.

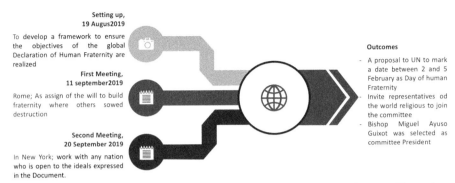

Setting up, 19 Augus2019
To develop a framework to ensure the objectives of the global Declaration of Human Fraternity are realized

First Meeting, 11 september2019
Rome; As assign of the will to build fraternity where others sowed destruction

Second Meeting, 20 September 2019
In New York; work with any nation who is open to the ideals expressed in the Document.

Outcomes
- A proposal to UN to mark a date between 2 and 5 February as Day of human Fraternity
- Invite representatives od the world religious to join the committee
- Bishop Miguel Ayuso Guixot was selected as committee President

Figure 1.11   Summary of the achievements of the Higher Committee of Human Fraternity.

<u>The responsibility of the Higher Committee of Human Fraternity</u>
The members of the higher committee are responsible for:

- establishing the groundwork for its future activities, as it is in its early stages of its formation;
- acting on the aspirations outlined in the Document on Human Fraternity;
- meeting with religious leaders, heads of international organizations, and others across the world;
- supporting and spreading the values of mutual respect and peaceful coexistence;
- providing counsel on a variety of initiatives, including the Abrahamic Family House to be built in Abu Dhabi;
- including leaders of other faiths, denominations, and beliefs;
- undertaking complex challenges facing communities of all faiths, with an approach of openness, learning, and dialogue.

Summary of the higher committee achievements is presented in Figure 1.11.

1.5 The Document of Human Fraternity

Introduction:
Faith leads a believer to see in the other a brother or sister to be supported and loved. Through faith in God, who has created the universe, creatures, and all human beings (equal on account of his mercy), believers are called to express this human fraternity by safeguarding creation and the entire universe and supporting all persons, especially the poorest and those most in need.

This transcendental value served as the starting point for several meetings characterized by a friendly and fraternal atmosphere where we shared the joys, sorrows, and problems of our contemporary world. We did this by considering scientific and technical progress, therapeutic achievements, the digital era, the mass media, and communications. We reflected also on the level of poverty, conflict, and suffering of so many brothers and sisters in different parts of the world as a consequence of the arms race, social injustice, corruption, inequality, moral decline, terrorism, discrimination, extremism, and many other causes.

From our fraternal and open discussions, and from the meeting that expressed profound hope in a bright future for all human beings, the idea of this *Document on Human Fraternity* was conceived. It is a text that has been given honest and serious thought so as to be a joint declaration of good and heartfelt aspirations. It is a document that invites all persons who have faith in God and faith in *human fraternity* to unite and work together so that it may serve as a guide for future generations to advance a culture of mutual respect in the awareness of the great divine grace that makes all human beings brothers and sisters.

Document:

In the name of God who has created all human beings equal in rights, duties, and dignity, and who has called them to live together as brothers and sisters, to fill the earth and make known the values of goodness, love, and peace.

In the name of innocent human life that God has forbidden to kill, affirming that whoever kills a person is like one who kills the whole of humanity, and that whoever saves a person is like one who saves the whole of humanity.

In the name of the poor, the destitute, the marginalized, and those most in need whom God has commanded us to help as a duty required of all persons, especially the wealthy and of means.

In the name of orphans, widows, refugees, and those exiled from their homes and their countries.

In the name of all victims of wars, persecution, and injustice.

In the name of the weak, those who live in fear, prisoners of war, and those tortured in any part of the world, without distinction.

In the name of people who have lost their security, peace, and the possibility of living together, becoming victims of destruction, calamity, and war.

In the name of *human fraternity* that embraces all human beings, unites them, and renders them equal.

In the name of this *fraternity* torn apart by policies of extremism and division, by systems of unrestrained profit, or by hateful ideological tendencies that manipulate the actions and the future of men and women.

In the name of freedom, which God has given to all human beings creating them free and distinguishing them by this gift.

In the name of justice and mercy, the foundations of prosperity and the cornerstone of faith.

In the name of all persons of good will present in every part of the world.

In the name of God and of everything stated thus far; Al-Azhar Al-Sharif and the Muslims of the East and West, together with the Catholic Church and the Catholics of the East and West, declare the adoption of a culture of dialogue as the path; mutual cooperation as the code of conduct; reciprocal understanding as the method and standard.

We, who believe in God and in the final meeting with Him and His judgment, on the basis of our religious and moral responsibility, and through this Document, call upon ourselves, upon the leaders of the world, as well as the architects of international policy and world economy to work strenuously to spread the culture of tolerance and of living together in peace; to intervene at the earliest opportunity to stop the shedding of innocent blood and bring an end to wars, conflicts, environmental decay and the moral and cultural decline that the world is presently experiencing.

We call upon intellectuals, philosophers, religious figures, artists, media professionals, and men and women of culture in every part of the world to rediscover the values of peace, justice, goodness, beauty, human fraternity, and coexistence in order to confirm the importance of these values as anchors of salvation for all, and to promote them everywhere.

This Declaration, setting out from a profound consideration of our contemporary reality, valuing its successes and in solidarity with its suffering, disasters, and calamities, believes firmly that among the most important causes of the crises of the modern world are a desensitized human conscience, a distancing from religious values and a prevailing individualism accompanied by materialistic philosophies that deify the human person and introduce worldly and material values in place of supreme and transcendental principles.

While recognizing the positive steps taken by our modern civilization in the fields of science, technology, medicine, industry, and welfare, especially in developed countries, we wish to emphasize that, associated with such historic advancements, great and valued as they are, there exists both a moral deterioration that influences international action and a weakening of spiritual values and responsibility. All this contributes to a general feeling of frustration, isolation, and desperation leading many to fall either into a vortex of atheistic, agnostic, or religious extremism, or into blind and fanatic extremism, which ultimately encourage forms of dependency and individual or collective self-destruction.

History shows that religious extremism, national extremism, and also intolerance have produced in the world, be it in the East or West, what might be referred to as signs of a "third world war being fought piecemeal." In several parts of the world and in many tragic circumstances, these signs have begun to be painfully apparent, as in those situations where the precise number of victims, widows, and orphans is unknown. We see, in addition, other regions preparing to become theaters of new conflicts, with outbreaks of tension and a build-up of arms and ammunition, and all this in a global context overshadowed by uncertainty, disillusionment, fear of the future, and controlled by narrow-minded economic interests.

We likewise affirm that major political crises, situations of injustice, and lack of equitable distribution of natural resources — which only a rich minority benefit from, to the detriment of the majority of the peoples of the earth — have generated, and continue to generate, vast numbers of poor, infirm, and deceased persons. This leads to catastrophic crises that various countries have fallen victim to despite their natural resources and the resourcefulness of young people which characterize these nations. In the face of such crises that result in the deaths of millions of children — wasted away from poverty and hunger — there is an unacceptable silence on the international level.

It is clear in this context how the family as the fundamental nucleus of society and humanity is essential in bringing children into the world, raising them, educating them, and providing them with solid moral formation and domestic security. To attack the institution of the family, to regard it with contempt, or to doubt its important role is one of the most threatening evils of our era.

We affirm also the importance of awakening religious awareness and the need to revive this awareness in the hearts of new generations through sound education and an adherence to moral values and upright religious teachings. In this way, we can confront tendencies that are individualistic, selfish, and conflicting and also address radicalism and blind extremism in all its forms and expressions.

The first and most important aim of religions is to believe in God, to honor Him and to invite all men and women to believe that this universe depends on a God who governs it. He is the Creator who has formed us with His divine wisdom and has granted us the gift of life to protect it. It is a gift that no one has the right to take away, threaten, or manipulate to suit oneself. Indeed, everyone must safeguard this gift of life from its beginning up to its natural end. We therefore condemn all those practices that are a threat to life such as genocide, acts of terrorism, forced displacement, human trafficking, abortion, and euthanasia. We likewise condemn the policies that promote these practices.

Moreover, we resolutely declare that religions must never incite war, hateful attitudes, hostility, and extremism, nor must they incite violence or the shedding of blood. These tragic realities are the consequence of a deviation from religious teachings. They result from a political manipulation of religions and from interpretations made by religious groups who, in the course of history, have taken advantage of the power of religious sentiment in the hearts of men and women in order to make them act in a way that has nothing to do with the truth of religion. This is done for the purpose of achieving objectives that are political, economic, worldly, and short-sighted. We thus call upon all concerned to stop using religions to incite hatred, violence, extremism, and blind fanaticism, and to refrain from using the name of God to justify acts of murder, exile, terrorism, and oppression. We ask this on the basis of our common belief in God who did not create men and women to be killed or to fight one another, nor to be tortured or humiliated in their lives and circumstances. God, the Almighty, has no need to be defended by anyone and does not want His name to be used to terrorize people.

This Document, in accordance with previous international documents that have emphasized the importance of the role of religions in the construction of world peace, upholds the following.

- The firm conviction that authentic teachings of religions invite us to remain rooted in the values of peace; to defend the values of mutual understanding, *human fraternity*, and harmonious coexistence; to re-establish wisdom, justice, and love; and to reawaken religious awareness among young people so that future generations may be protected from the realm of materialistic thinking and from dangerous policies of unbridled greed and indifference that are based on the law of force and not on the force of law.

- Freedom is a right of every person: each individual enjoys the freedom of belief, thought, expression, and action. The pluralism and the diversity of religions, color, sex, race, and language are willed by God in His wisdom, through which He created human beings. This divine wisdom is the source from which the right to freedom of belief and the freedom to be different derives. Therefore, the fact that people are forced to adhere to a certain religion or culture must be rejected, as too the imposition of a cultural way of life that others do not accept.
- Justice based on mercy is the path to follow in order to achieve a dignified life to which every human being has a right;
- Dialogue, understanding, and the widespread promotion of a culture of tolerance, acceptance of others and of living together peacefully would contribute significantly to reducing many economic, social, political, and environmental problems that weigh so heavily on a large part of humanity.
- Dialogue among believers means coming together in the vast space of spiritual, human, and shared social values and, from here, transmitting the highest moral virtues that religions aim for. It also means avoiding unproductive discussions.
- The protection of places of worship — synagogues, churches, and mosques — is a duty guaranteed by religions, human values, laws, and international agreements. Every attempt to attack places of worship or threaten them by violent assaults, bombings, or destruction is a deviation from the teachings of religions as well as a clear violation of international law.
- Terrorism is deplorable and threatens the security of people, be it in the East or the West, the North or the South, and disseminates panic, terror, and pessimism, but this is not due to religion, even when terrorists instrumentalize it. It is due, rather, to an accumulation of incorrect interpretations of religious texts and to policies linked to hunger, poverty, injustice, oppression, and pride. This is why it is so necessary to stop supporting terrorist movements fuelled by financing, the provision of weapons and strategy, and by attempts to justify these movements even using the media. All these must be regarded as international crimes that threaten security and world peace. Such terrorism must be condemned in all its forms and expressions.
- The concept of *citizenship* is based on the equality of rights and duties, under which all enjoy justice. It is therefore crucial to establish in our societies the concept of *full citizenship* and reject the discriminatory

use of the term *minorities* which engenders feelings of isolation and inferiority. Its misuse paves the way for hostility and discord; it undoes any successes and takes away the religious and civil rights of some citizens who are thus discriminated against.

- Good relations between East and West are indisputably necessary for both. They must not be neglected so that each can be enriched by the other's culture through fruitful exchange and dialogue. The West can discover in the East remedies for those spiritual and religious maladies that are caused by a prevailing materialism. And the East can find in the West many elements that can help free it from weakness, division, conflict, and scientific, technical, and cultural decline. It is important to pay attention to religious, cultural, and historical differences that are a vital component in shaping the character, culture, and civilization of the East. It is likewise important to reinforce the bond of fundamental human rights in order to help ensure a dignified life for all the men and women of East and West, avoiding the politics of double standards.
- It is an essential requirement to recognize the right of women to education and employment, and to recognize their freedom to exercise their own political rights. Moreover, efforts must be made to free women from historical and social conditioning that runs contrary to the principles of their faith and dignity. It is also necessary to protect women from sexual exploitation and from being treated as merchandise or objects of pleasure or financial gain. Accordingly, an end must be brought to all those inhuman and vulgar practices that denigrate the dignity of women. Efforts must be made to modify those laws that prevent women from fully enjoying their rights.
- The protection of the fundamental rights of children to grow up in a family environment, to receive nutrition, education, and support, are duties of the family and society. Such duties must be guaranteed and protected so that they are not overlooked or denied to any child in any part of the world. All those practices that violate the dignity and rights of children must be denounced. It is equally important to be vigilant against the dangers that they are exposed to, particularly in the digital world, and to consider as a crime the trafficking of their innocence and all violations of their youth.
- The protection of the rights of the elderly, the weak, the disabled, and the oppressed is a religious and social obligation that must be guaranteed and defended through strict legislation and the implementation of the relevant international agreements.

To this end, by mutual cooperation, the Catholic Church and Al-Azhar announce and pledge to convey this Document to authorities, influential leaders, persons of religion all over the world, appropriate regional and international organizations, organizations within civil society, religious institutions, and leading thinkers. They further pledge to make known the principles contained in this Declaration at all regional and international levels, while requesting that these principles be translated into policies, decisions, legislative texts, courses of study, and materials to be circulated.

Al-Azhar and the Catholic Church ask that this Document become the object of research and reflection in all schools, universities, and institutes of formation, thus helping to educate new generations to bring goodness and peace to others, and to be defenders everywhere of the rights of the oppressed and of the least of our brothers and sisters.

In conclusion, our aspiration is the following.

- This Declaration may constitute an invitation to reconciliation and fraternity among all believers, indeed among believers and non-believers, and among all people of good will.
- This Declaration may be an appeal to every upright conscience that rejects deplorable violence and blind extremism; an appeal to those who cherish the values of tolerance and fraternity that are promoted and encouraged by religions.
- This Declaration may be a witness to the greatness of faith in God that unites divided hearts and elevates the human soul.
- This Declaration may be a sign of the closeness between East and West, between North and South, and between all who believe that God has created us to understand one another, cooperate with one another, and live as brothers and sisters who love one another.
- This is what we hope and seek to achieve with the aim of finding a universal peace that all can enjoy in this life.

2

Peace, Conflict, and War: The Role of Language and Languages

Prof. Timothy Reagan

University of Maine

Language makes us human. Whatever we do, language is central to our lives, and the use of language underpins the study of every other discipline. Understanding language gives us insight into ourselves and a tool for the investigation of the rest of the universe. Proposing marriage, opposing globalization, composing a speech, all require the use of language; to buy a meal or sell a car involves communication, which is made possible by language; to be without language — as an infant, a foreigner or a stroke victim — is to be at a devastating disadvantage. Martians and dolphins, bonobos and bees, may be just as intelligent, cute, adept at social organization, and morally worthwhile, but they do not share our language, they do not speak "human" (Smith, 2002, p. 3).

The leading linguist of the 20th century, Noam Chomsky (1972), once commented that, "When we study human language, we are approaching what some might call the 'human essence', the distinctive qualities of mind that are, so far as we know, unique to man" (Fromkin, Rodman & Hyams, 2003, p. 3). If language is unique to humanity, conflict is most certainly not. War is perhaps uniquely human (though there is evidence that chimpanzees also engage in war-like behavior), but this is certainly nothing to boast about. If we are more successfully war-like than other species, this is due to our mastery of the technology rather than to any particular cognitive, let alone moral superiority. In fact, war is an especially puzzling part of the human experience. As Dale Copeland has argued, "Since Thucydides, the puzzle of major war has been one of the most important but intractable questions in the study of international relations" (2000, p. 1) — and not simply international relations, but in trying to understand human beings more generally.

The purpose of this chapter is to explore the complex relationships among peace, conflict, and war on the one hand, and language and language diversity on the other. Although these relationships are important, the fundamental argument presented here will be that, in spite of many historical and contemporary claiming the contrary, linguistic differences are not causally related to other kinds of conflict, but other kinds of conflict are often manifested in linguistic conflict.

Diversity is a core characteristic of the human experience, as virtually any introductory anthropology textbook will suggest (e.g., Kottak, 2014; Muckle & Lubelle de González, 2016). We differ in a wide host of ways; human beings live in radically different kinds of societies: embodying an array of political and economic systems, structuring their families and kinship systems in almost countless manners, socializing and educating their children in an incredibly mix of different ways, practicing a wide assortment of religions and spiritual systems, utilizing multitudinous kinds of technology of varying degrees of sophistication, dressing themselves in a plentiful range of different garments and apparel, surviving in virtually all of the many climates on the planet, and otherwise coping with meeting their daily needs in a variety of ways. We also use an extensive number of languages, which vary in an incredible abundance of ways (e.g., McGregor, 2015; Radford, Atkinson, Britain, Clahsen & Spencer, 2009; Yule, 2017). Nor, it is important to keep in mind, are either human culture or human language ever static; both are in a perpetual process of change. As David Pharies has commented:

> "Human culture is constantly changing in every way: in the way people dress or wear their hair; in the technologies they use; in their political, religious, and educational institutions; in the way they treat children and animals; in what and how much they eat; in the way the sexes relate to each other. Language can be characterized as the ultimate manifestation of human culture. It represents the foundation, in practical terms, of all other cultural elements, since it is the instrument through which is conveyed the entire body of knowledge that constitutes our customs, laws, and concept of human life. Perhaps because language is so omnipresent in our lives, the subtle yet infinite series of changes that it undergoes are sometimes difficult to perceive." (2007, p. 1)

Most linguists estimate that there are somewhere between 6500 and 7000 languages currently spoken on our world.[1] Although certainly an impressive number, such estimates take into account only a tiny percentage of the total

number of human languages that must have existed since homo sapiens —
or, more technically accurate, homo sapiens sapiens — first emerged as a
distinctive species at some point around 250,000 years ago, when all of the
modern populations of human beings diverged from a common ancestor, a
time known as the "Mitochondrial Eve." Human beings, for most of our
existence as a species, have lived in extremely small groups, making the prob-
ability for the existence of an extremely large number of different languages
quite high (Barnard, 2016, p. 5). One estimate, for instance, has suggested
that as recently as 8000 BCE, there were around 20,000 different languages
spoken by human beings. As James Hurford has suggested:

> "It is likely that in prehistory, even though the human population
> was much smaller, the number of languages was greater. The num-
> ber of different languages that have ever existed is far greater than
> the number we can count now. To grasp this, we have to abandon
> the notion of global languages like English, Chinese, and Arabic,
> spoken by millions." (2014, p. 16)

There is a great deal that we do not know, and will never know, about the
origins of human language. We do not know when language first began
to be used by our ancestors, nor do we know where it first emerged, or
whether it began as a common, single, proto-language somewhere and then
spread or if, instead, it started as a number of distinct languages used by
different groups in different locations — comparable to parallel evolution
in. We do not know whether language emerged suddenly in a more or less
complete linguistic form such as the languages that exist today — referred to
as the "discontinuity hypothesis" — or whether it was the result of a much
longer process of evolution and development — known as the "continuity
hypothesis." Although some anthropological linguists have argued that what
we could consider "human language" to have emerged far earlier (Barnard,
2016, pp. 35–37), most linguists believe that while our ancestors had the
cognitive and physiological capacity for language earlier, they only began
using what we would call "language" around 100,000–50,000 years ago,
which is admittedly quite a range. Even this extremely conservative estimate
suggests that, given rates of normal language change, there have been tens of
thousands of distinct human languages in our prehistory. To be sure, this is
speculation since for most of our existence as a species, there are no records
of any type of the languages used by human beings.

   The first documented human languages date back some five thousand
years to Mesopotamia, where we first have written records of Sumerian,

an isolated language containing elements of which were maintained in a complex linguistic and cultural bilingual symbiosis, which later developed as the Akkadians gradually replaced the Sumerians as a leading imperial power (Cooper, 1973; Edzard, 2003; Gianto, 1999; Huehnergard, 2011; Michalowski, 1996). Parts of the Gilgamesh epic cycle first appear in Sumerian, although the more complete version is found in Akkadian. Akkadian was an East Semitic language, linguistically unrelated to Sumerian but written in the same cuneiform script and with heavy borrowings from Sumerian.[2] Akkadian, in turn, was gradually replaced by Assyrian, which functioned as a lingua franca throughout much of the ancient Near East, declining as first Aramaic, and later koine Greek, the primary language of the New Testament, took over this function (Siegel, 2009). My point in reciting this history of early Mesopotamian languages is simply that language diversity, coupled with both social and individual bilingualism, and even multilingualism, has long been characteristic of human societies. Societies such as that of the United States and, indeed, most Anglophone societies, in which one language overwhelmingly dominates linguistic communication and in which both individual and social monolingualism are considered to be the norm, are in fact extremely atypical in the human experience.

In considering language and language diversity, it has been suggested that the use of different languages is frequently the cause of conflict between different groups. Diarmait Mac Giolla Chríost, for instance, has commented that "[t]he idea of language, in part, and especially of language in conflict, resides in a complexity of relationships between self-identification group cohesion and world-view" (2003, p. 9). One example of this can be seen in the case of Lazar Ludwig Zamenhof, who grew up in the town of Bialystok, which was then in the part of Poland that was under the control of the Russian Empire (see Garvía, 2015, pp. 60–64; Korzhenkov, 2010; Okrent, 2009, pp. 94–95; Schor, 2016). Zamenhof witnessed, firsthand, the tensions and conflicts between the different groups who lived in Bialystok. As he explained in a letter to Nikolai Borovko in 1895:

> "In Bialystok the population consisted of four different elements: Russians, Poles, Germans and Jews. Each of these elements spoke a separate language and had hostile relations with the other elements. In that city, more than anywhere, a sensitive person might feel the heavy sadness of the diversity of languages and become convinced at every step that it is the only, or at least the primary force

which divides the human family into enemy parts." (Okrent, 2009, pp. 94–95)

Zamenhof's solution to the conflict between ethnic, national, and linguistic groups was the creation of a language that would be politically and ethically neutral, and it was this motivation that led to the creation of the international auxiliary language Esperanto (see Janton, 1993; Nuessel, 2000).[3] Zamenhof was absolutely correct in his observations about the conflicts and tensions between and among the different groups in Bialystok and, indeed, throughout the Pale of Settlement — the area in which Jews were allowed to live and outside of which they were generally forbidden to live on anything but a temporary basis — but his solution has not proven to be a particularly successful one.[4]

Language diversity has typically been related to several other social, political, and economic features that have important implications here. The first of these tendencies is that of language contact and, as a consequence, language dominance. Although not universal, most often, a conquering or ruling group imposes its language on those whom it has conquered or over whom it rules. Sometimes this takes place informally, while in others, it is the result of a deliberate policy. It often results in language shift, as well as phonological, morphological, and syntactic changes in all of the languages involved — although most obviously in lexical changes (Bybee, 2015). Such changes are frequently the result of the development of social and individual bilingualism and multilingualism in the short and medium terms, and not only language change but also language replacement; language emergence, as in cases of creolization; and language endangerment and death in the longer run (e.g., Austin & Sallabank, 2012; Evans, 2010; Grenoble & Whaley, 1998; Hagège, 2000; Jones, 2015; Nettle & Romaine, 2000; Thomason, 2015). In the contemporary era, two phenomena have become increasingly serious. The first of these phenomena is the growing dominance of a small number of Languages of Wider Communication (LWCs), and especially of English, in the world. The hegemony of English as a world language is largely unmatched in human history. Whether we are talking about diplomacy, commerce, scientific publications and presentations, entertainment, newspaper publishing, communication on the Internet, or pop culture, the English language plays the central role on the world stage — leading, of course, to very reasonable concerns about linguistic imperialism (Phillipson, 1992, 1997, 2006, 2008, 2009). The second phenomenon of the modern era is the growing threat of language endangerment or language death. Michael Krauss (1992) has suggested that within the next 150 years, only somewhere between 300 and 600 of the languages currently used in the world will remain, a loss in the neighborhood of some 90% of the present 6500 languages (Evans, 2010; Grenoble & Whaley, 1998; Hagège, 2000; Jones, 2015; Nettle & Romaine, 2000).

The dominance of particular languages in different societies commonly leads to linguistic inequality. Although it is certainly possible to imagine a human society in which two (or more) languages coexist on a basis of equality in all domains, such societies have been rare. The sociolinguistic norm in linguistically diverse societies entails the development of diglossic situations in which two language varieties are present in a single language community.[5] In such societies, one of the varieties is usually the L (low) linguistic variety, which is used as the daily vernacular language, and the H (high) variety, which is used in specific settings, such as literature, education, government, and so on. Even in monolingual societies, there are different varieties of the common language, and the varieties are virtually never perceived to be (or treated as) equal linguistic varieties; one's language or language variety is closely tied not only to identity but even more to status and power (Trudgill, 2016; Wardhaugh, 1999; for the specific case of the United States, see Lippi-Green, 2012). As the country song "Good Ole Boys Like Me" by Don Williams first released in 1979, it suggested that part of success in modern America is learning "to talk like the man on the six o'clock news."

Not only are different linguistic varieties tied to different kinds of identity, and carry with them different degrees of status and power, but they are also inevitably judged and evaluated by members of the society. In the contemporary USA, in addition to the H variety of Standard American English, there are a number of L varieties of the language, including African American English (Baugh, 2000; Green, 2002, 2011; McWhorter, 1998; Morgan, 2002; Mufwene, Rickford, Bailey & Baugh, 1998; Rickford, 2006; Rickford & Rickford, 2000), varieties of Southern English (Lippi-Green 2012, pp. 214–234), "Spanglish" (González Echevarría, 1997; Morales, 2002; Otheguy & Stern, 2011; Sánchez-Muñoz, 2013), and so on. Such varieties play a key role in maintaining social, economic, and political discrimination in the United States. In this regard, Leah Zuidema has noted, "Linguistic prejudice is one of the few 'acceptable' American prejudices. In polite society, we don't allow jokes that we consider racist or sexist, and we are careful not to disparage a person's religious beliefs. Language is another matter" (2005, p. 686). Not only do we make judgments about the language variety an individual speaks, but we feel perfectly comfortable in expecting speakers of L varieties of

American English to replace their language variety with a more acceptable H variety. As Rosina Lippi-Green has observed, "We do not, cannot under our laws, ask people to change the color of their skin, their religion, their gender, but we regularly demand of people that they suppress or deny the most effective way they have of situating themselves socially in the world" (1997, p. 63).

These facets of linguistic diversity have often been correlated with significant levels of conflict in societies throughout human history, at least in part because of the centrality of language to both individual and group identity. Examples of such conflict, which, of course, are related not simply to language but also to other political, social, and economic factors, abound. David Laitin has noted that a Tower of Babel in a single country, in which groups of people speak radically different languages, is all too often portrayed as incendiary. Selig Harrison wrote ominously about the "dangerous decades" that India would face because of its conflicts over language. Popular representations of language conflicts in Belgium, Quebec, and Catalonia suggest that cultural issues of this sort unleash irrational passions, leading otherwise sober people away from the realm of civic engagement.

The recent independence referendum in Catalonia is one clear example (Moreno, 2008; Strubell & Boix-Fuster, 2011; Woolard, 2003, 2013), but others can be found in virtually every part of the world. The language reform which took place in Turkey, following the end of the Ottoman Empire and the establishment of the Turkish Republic, was, for instance, concerned in large part with both orthographic reform, as the Arabic script was replaced with the Latin script, and with efforts to "purify" the Turkish language by eliminating foreign borrowings from Arabic and Persian (e.g., Boeschoten, 1997; Dogançay-Aktuna, 1995; Lewis, 1999; Perry, 1985). At the same time, by attempting to promote one sort of ethnolinguistic nationalism, the Turkish regime effectively disenfranchised other groups, leading in part to the current tensions with the Kurds (Hassanpour, Skutnabb-Kangas & Chyet, 1996; Yavuz, 1998; Zeydanlioglu, 2012). After the collapse of the Soviet Union, almost all of the newly independent states, excluding Russia itself, as well as Belarus and, for a time, Ukraine, rapidly implemented changes in official language legislation that either drastically reduced or eliminated altogether the role of Russian, leading in turn to language conflicts between native speakers of Russian and others in these countries (Brubaker, 2011; Fierman, 2005; Marshall, 2002; Ozolins, 2003; Pavlenko, 2008a, 2008b).

If language conflict is sometimes the result of language difference, it is, nevertheless, important to note that merely sharing a common language in no way ensures a lack of conflict. By the end of the 19th century, Ireland was an overwhelmingly monolingual society in which English was the common ver- nacular, but this in no way minimized the fight for home rule and ultimately independence in the country (Buachalla, 1984; Hindley, 1991; Walsh, 2012). In the history of the United States, both the American Revolutionaries and the Tories were English speakers, as, indeed, were soldiers and citizens of both the Union and the Confederacy during the Civil War. More recently, the fact that Serbian and Croatian are arguably simply two varieties of a common South Slavic language, albeit written in two alphabets, in no way prevented the collapse and dismemberment of Yugoslavia, nor did it in any way minimize the horrors of that experience (Magas, 1993; Radan, 2002; Ramet, 2005). In short, it appears to be fairly clear that a shared, common language has little, if any, impact in terms of promoting a common, shared identity, nor even in necessarily promoting good relations between groups if other factors, such as ethnicity, religion, economics, politics, ideology, and so on, are seen as more important.

So, where does this bring us? What has been argued here is that while language is indeed a central feature of our humanity, and although it plays an incredibly important part in our daily lives, its power is tied closely to a variety of other factors, and that while it can and does often play a role in both creating and remediating conflicts between groups on its own, its impact is far more limited. As Peter Nelde observed some years ago:

> "The height of a political language conflict is reached when all conflict factors are combined in a single symbol, language, and quarrels and struggles in very different areas [politics, economics, administration, education] appear under the heading language con- flict. In such cases, politicians and economic leaders also operate on the assumption of language conflict, disregarding the actual underlying causes, and thus continue to feed "from above" the conflict that has arisen "from below," with the result that language assumes much more importance than it had at the outset of the conflict. This language-oriented "surface symptom" then obscures the more deeply rooted, suppressed "deeper causes" [social and economic problems]." (1987, p. 35)

There can be no doubt that in many cases in which we find conflict, intol- erance, and insensitivity, these are reflected in both attitudes and actions concerned with language. Examples of this phenomenon abound, whether in the denial of basic services to speakers of a language, the lack of education to children in their mother tongue, access to political and economic power, or in a host of other ways — all of which constitute, in one way or another, the violation of the linguistic human rights of speakers of marginalized languages (Dunbar, 2001; Faingold, 2018; Grin, 2005; Hornberger, 1998; May, 2003, 2006, 2012; Paulston, 2003; Skutnabb-Kangas & Phillipson, in conjunction with Rannut, 1995; Stroud, 2010). Addressing such violations of fundamental linguistic human rights will not, in the vast majority of instances, completely resolve or eliminate the underlying conflict, but it is, at the very least, a necessary condition for doing so and an important step both symbolically and practically.

Notes

Ethnologue: Languages of the world is published on an annual basis by SIL International and is widely considered to be a fairly standard reference work on the identified languages of the world. The 20th edition of Ethnologue, published in 2017, included 7099 languages.

The cuneiform script of Akkadian, including borrowings from Sumerian, was later adopted as the orthography of the Hittite language, spoken in north-central Anatolia in what is today Turkey. Hittite is the oldest attested Indo-European language (Hoffner & Melchert, 2008; Jasanoff, 2003; van den Hout, 2011).

Although there is on-going and quite good, linguistic research being conducted on Esperanto, the best general overview of the grammar of Esperanto grammar is Wennergren (2005).

In fact, determining the success or failure of Esperanto is no easy matter. Although the language never met Zamenhof's own hopes, it is, by far, the most successful of all of the international auxiliary language projects, and has a well-established international speaker community.

The classic description of diglossia is provided in Ferguson (1959), but there is now an extensive research literature on the topic.

References

Austin, P. & Sallabank, J. (Eds.). (2012). The Cambridge handbook of endangered languages. Cambridge: Cambridge University Press.

Barnard, A. (2017). Language in prehistory. Cambridge: Cambridge University Press.

Baugh, J. (2000). Beyond Ebonics: Linguistic pride and racial prejudice. Oxford: Oxford University Press.

Boeschoten, H. (1997). The Turkish language reform forced into stagnation. In M. Clyde (Ed.), Undoing and redoing corpus planning (pp. 357–383). Berlin: Mouton de Gruter.

Brubaker, R. (2011). Nationalizing states revisited: Projects and processes of nationalization in post-Soviet states. Ethnic and Racial Studies, 34(11): 1785–1814.

Buachalla, A. (1984). Education policy and the role of the Irish language from 1831 to 1981. European Journal of Education, 19(1): 75–92.

Bybee, J. (2015). Language change. Cambridge: Cambridge University Press.

Chomsky, N. (1972). Language and mind (enlarged ed.). New York: Harcourt Brace Jovanovich.

Cooper, J. (1973). Sumerian and Akkadian in Sumer and Akkad. Orientalia (Nova Series), 42: 239–246.

Copeland, D. (2000). The origins of major war. Ithaca, NY: Cornell University Press.

Dogançay-Aktuna, S. (1995). An evaluation of the Turkish language reform after 60 years. Language Problems and Language Planning, 19(3): 221–249.

Dunbar, R. (2001). Minority language rights in international law. International and Comparative Law Quarterly, 50(1): 90–120.

Edzard, D. (2003). Sumerian grammar. Leiden: Brill.

Evans, N. (2010). Dying words: Endangered languages and what they have to tell us. Malden, MA: Wiley-Blackwell.

Faingold, E. (2018). Language rights and the law in the United States and its territories. Lanham, MD: Lexington Books.

Ferguson, C. (1959). Diglossia. Word, 15(2): 325–340.

Fierman, W. (2005). Language and education in post-Soviet Kazakhstan: Kazakh-medium instruction in urban schools. The Russian Review, 65(1): 98–116.

Fromkin, V., Rodman, R. & Hyams, N. (2003). An introduction to language (5th ed.). Cambridge: Cambridge University Press.

Garvía, R. (2015). Esperanto and its rivals: The struggle for an international language. Philadelphia: University of Pennsylvania Press.

Gianto, A. (1999). Amarna Akkadian as a contact language. In K. van Lergerghe and G. Voet (Eds.), Languages and cultures in contact at the crossroads of civilizations in the Syro-Mesopotamian realm: Proceedings of the 42nd RAI (pp. 123–132). Leuven: Uitgeverij Peeters en Departement Oosterse Studies.

González Echevarría, R. (1997). Is "Spanglish" a language? The New York Times, p. A–29 (March 28).

Green, L. (2002). African American English: A linguistic introduction. Cambridge: Cambridge University Press.

Green, L. (2011). Language and the African American child. Cambridge: Cambridge University Press.

Grenoble, L. & Whaley, L. (Eds.). (1998). Endangered languages: Current issues and future prospects. Cambridge: Cambridge University Press.

Grin, F. (2005). Linguistic human rights as a source of policy guidelines: A critical assessment. Journal of Sociolinguistics, 9(3): 448–460.

Hagège, C. (2000). Halte à la mort des langues. Paris: Editions Odile Jacob.

Hassanpour, Am., Skutnabb-Kangas, T. & Chyet, M. (1996). The non-education of Kurds: A Kurdish perspective. International Review of Education, 42(4): 367–379.

Hindley, R. (1991). The death of the Irish language: A qualified obituary. London: Routledge.

Hoffner, H. & Melchert, H. (2008). A grammar of the Hittite language, 2 vols. Winona: Eisenbrauns.

Hornberger, N. (1998). Language policy, language education, language rights: Indigenous, immigrant, and international perspectives. Language in Society, 27(4): 439–458.

Huehnergard, J. (2011). A grammar of Akkadian (3rd ed.). Winona Lake, IN: Eisenbrauns.

Hurford, J. (2014). The origins of language: A slim guide. Oxford: Oxford University Press.

Janton, P. (1993). Esperanto: Language, literature, and community. Albany: State University of New York Press.

Jasanoff, J. (2003). Hittite and the Indo-European verb. Oxford: Oxford University Press.

Jones, M. (Ed.). (2015). Policy and planning for endangered languages. Cambridge: Cambridge University Press.

Korzhenkov, A. (2010). Zamenhof: The life, works and ideas of the author of Esperanto. New York: Mondial, in cooperation with the Universala Esperanto Asocio.

Kottak, C. (2014). Window on humanity: A concise introduction to anthropology (9th ed.). New York: McGraw-Hill Education.

Krauss, M. (1992). The world's languages in crisis. Language, 68(1): 4–10.

Lewis, G. (1999). The Turkish language reform: A catastrophic success. Oxford: Oxford University Press.

Lippi-Green, R. (1997). English with an accent: Language, ideology and discrimination in the United States. London: Routledge.

Lippi-Green, R. (2012). English with an accent: Language, ideology, and discrimination in the United States (2nd ed.). London: Routledge.

Mac Giolla Chríost, D. (2003). Language, identity and conflict: A comparative study of language in ethnic conflict in Europe and Eurasia. London: Routledge.

Magas, B. (1993). The destruction of Yugoslavia: Tracking the break-up 1980–92. London: Verso.

Marshall, C. (2002). Post-Soviet language policy and the language utilization patterns of Kyivan youth. Language Policy, 1(3): 237–260.

May, S. (2003). Rearticulating the case for minority language rights. Current Issues in Language Planning, 4(2): 95–125.

May, S. (2006). Language policy and minority rights. In R. Ricento (Ed.), An introduction to language policy: Theory and method (pp. 255–272). Oxford: Blackwell.

May, S. (2012). Language and minority rights: Ethnicity, nationalism and the politics of language. New York and London: Routledge.

McGregor, W. (2015). Linguistics: An introduction (2nd ed.). London: Bloomsbury Academic.

McWhorter, J. (1998). The word on the street: Fact and fable about American English. New York: Plenum.

Michalowski, P. (1996). Mesopotamian cuneiform. In P. Daniels and W. Bright (Ed.), The world's writing systems (pp. 33–57). Oxford: Oxford University Press.

Morales, E. (2002). Living in Spanglish: The search for Latino identity in America. New York: Macmillan.

Moreno, F. (2008). Language-in-education policies in the Catalan language area. AILA Review, 21(1): 31–48.

Morgan, M. (2002). Language, discourse and power in African American culture. Cambridge: Cambridge University Press.

Muckle, R. & Lubelle de González, L. (2016). Through the lens of anthropology: An introduction to human evolution and culture. Toronto: University of Toronto Press.

Mufwene, S., Rickford, J., Bailey, G. & Baugh, J. (Eds.). (1998). African American English: Structure, history and use. London: Routledge.

Nelde, P. (1987). Language contact means language conflict. In G. MacEoin, A. Ahlquist and D. Óh Aodha (Eds.), 3ʳᵈ International Conference on Minority Languages. Clevedon: Multilingual Matters.

Nettle, D. & Romaine, S. (2000). Vanishing voices: The extinction of the world's languages. Oxford: Oxford University Press.

Nuessel, F. (2000). The Esperanto language. New York: Legas.

Okrent, A. (2009). In the land of invented languages: Esperanto rock stars, Klingon poets, Loglan lovers, and the mad dreamers who tried to build a perfect language. New York: Spiegel & Grau.

Otheguy, R. & Stern, N. (2011). On so-called Spanglish. International Journal of Bilingualism, 15(1): 85–100.

Ozolins, U. (2003). The impact of European accession upon language policy in the Baltic states. Language Policy, 2(3): 217–238.

Paulston, C. (2003). Language policies and language rights. In C. Paulston and G. Tucker (Eds.), Sociolinguistics: The essential readings (pp. 472–483). Oxford: Blackwell.

Pavlenko, A. (Ed.). (2008a). Multilingualism in post-Soviet countries. Bristol: Multilingual Matters.

Pavlenko, A. (2008b). Multilingualism in post-Soviet countries: Language revival, language removal, and sociolinguistic theory. International Journal of Bilingual Education and Bilingualism, 11(3/4): 275–314.

Perry, J. (1985). Language reform in Turkey and Iran. International Journal of Middle East Studies, 17(3), 295–311.

Pharies, D. (2007). A brief history of the Spanish language. Chicago: University of Chicago Press.

Phillipson, R. (1992). Linguistic imperialism. Oxford: Oxford University Press.

Phillipson, R. (1997). Realities and myths of linguistic imperialism. Journal of Multilingual and Multicultural Development, 18(3): 238–248.

Phillipson, R. (2008). The linguistic imperialism of neoliberal empire. Critical Inquiry in Language Studies, 5(1): 1–43.

Phillipson, R. (2009). Linguistic imperialism continued. London: Routledge.

Radan, P. (2002). The break-up of Yugoslavia and international law. London: Routledge.

Radford, A., Atkinson, M., Britain, D., Clahsen, H. & Spencer, A. (2009). Linguistics: An introduction (2ⁿᵈ ed.). Cambridge: Cambridge University Press.

Ramet, S. (2005). Thinking about Yugoslavia: Scholarly debates about the Yugoslav breakup and the wars in Bosnia and Kosovo. Cambridge: Cambridge University Press.

Rickford, J. (2006). Linguistics, education, and the Ebonics firestorm. In S. Nero (Ed.), Dialects, Englishes, creoles, and education (pp. 71–92). Mahwah, NJ: Lawrence Erlbaum Associates.

Rickford, J. & Rickford, R. (2000). Spoken soul: The story of Black English. New York: John Wiley & Sons.

Sánchez-Muñoz, A. (2013). Who soy yo? The creative use of "Spanglish" to express a hybrid identity in chicana/o heritage language learners of Spanish. Hispania, 96(3): 440–441.

Schor, E. (2016). Bridge of words: Esperanto and the dream of a universal language. New York: Metropolitan Books.

Siegel, J. (2009). Introduction: Controversies in the study of koines and koineization. International Journal of the Sociology of Language, 99(1): 5–8.

Skutnabb-Kangas, T. & Phillipson, R. in conjunction with Rannut, M. (Eds.). (1995). Linguistic human rights. Berlin: Mouton de Gruyter.

Smith, N. (2002). Language, bananas and bonobos: Linguistic problems, puzzles and polemics. Oxford: Blackwell.

Stroud, C. (2010). African mother-tongue programmes and the politics of language: Linguistic citizenship vs. linguistic human rights. Journal of Multilingual and Multicultural Development, 22(4): 339–355

Strubell, M. & Boix-Fuster, E. (Eds.). (2011). Democratic policies for language revitalization: The case of Catalan. New York: Palgrave Macmillan.

Thomason, S. (2015). Endangered languages: An introduction. Cambridge: Cambridge University Press.

Trudgill, P. (2016). Dialect matters: Respecting vernacular language. Cambridge: Cambridge University Press.

van den Hout, T. (2011). The elements of Hittite. Cambridge: Cambridge University Press.

Walsh, J. (2012). Language policy and language governance: A case study of Irish language legislation. Language Policy, 11(4): 323–341.

Wardhaugh, R. (1999). Proper English: Myths and misunderstandings about language. Oxford: Blackwell.

Woolard, K. (2003). "We don't speak Catalan because we are marginalized": Ethnic and class meanings of language in Barcelona. In R. Blot (Ed.), Language and social identity (pp. 85–104). Westport, CT: Praeger.

Woolard, K. (2013). Is the personal political? Chronotopes and changing stances toward Catalan language and identity. International Journal of Bilingual Education and Bilingualism, 16(2): 210–224.

Yavuz, M. (1998). A preamble to the Kurdish question: The politics of Kurdish identity. Journal of Muslim Minority Affairs, 18(1): 9–18.

Yule, G. (2017). The study of language (6th ed.). Cambridge: Cambridge University Press.

Zeydanlioglu, W. (2012). Turkey's Kurdish language policy. International Journal of the Sociology of Language, 217: 99–125.

Zuidema, L. (2005). Myth education: Rationale and strategies for teaching against linguistic prejudice. Journal of Adolescent and Adult Literacy, 48(8): 666–676.

3

An Institutional Model for Tolerance and Peace Using a Formulaic Integration of Equity, Diversity, and Inclusion

Prof. David L. Everett

Hamline University

3.1 Introduction

Given the hyper-polarizing taking place within the society recently, there have been gentle overtures for tolerance and peace. These overtures typically take the form of either creating specific positions or learning opportunities that take the place of institutional apparatus or operate in parallel. Many institutions have established, by various names, an entity that is charged with weaving tolerance and peace objectives into existing equity, diversity, and inclusion goals, for example, Chief Diversity Officer and Director of Inclusive Excellence and Institutional Culture titles, just to name a few. The success of this approach and roles is mixed. This chapter proposes a road map that, while not guaranteeing success, increases the chance of genuine integration of tolerance and peace through new ways of thinking about and approaching equity, diversity, and inclusion.

In exploring genuine integration of tolerance and peace, institutions need to focus on essential components that contribute to leadership development, ownership characteristics, and partnership opportunities. They need to expand their thinking about the practical meaning of tolerance and peace and find ways to establish critical methods of engagement, exploration, and evaluation. Traditional approaches, such as affirmative action programs and diversity training workshops, have proven to be ineffective and even counterproductive as they tend to conflate individual and institutional implications. When the two are properly delineated and defined, however, the most effective approaches make strategic and synergistic use of competency, capacity, and community. Institutions in which constituents and stakeholders are equipped, accountable, and connected significantly improve efforts that can have a transformational advantage not easily offset by other institutional characteristics and social dynamics.

Institutions and systems throughout the world are undergoing substantial demographic change, with members of previously under-represented and unrepresented groups making up increasing proportions. This inclusion of members of minoritized groups is not, however, a comprehensive reflection of tolerance and peace — tolerance, in this sense, equates to *equity*, and peace equates to *inclusion*. Too often, institutions and systems have settled for the goal of diversity versus inclusion and equity, and thus have been agnostic about prescriptive-only culture dynamics. The evidence, in fact, suggests that not only do institutions and systems often reproduce rather than remedying patterns of marginalization, exclusion, and oppression, but also that substantial disparities remain between and among groups across a number of wellness indicators. Thus, those seeking to understand and address these patterns and disparities must do the following three things: examine current institutional realities, address the systemic nature of those realties, and make the connection between realities and culture.

Current realities highlight issues of tolerance and peace within our society — refusal to collaborate, unwillingness to engage different narratives, and the labeling of dissenting experiences as uninformed, to name but a few. These realities, however, upsetting and unfortunate, have created a tension of opportunity to discuss tolerance and peace in a more robust and substantive manner. The fundamental understanding of the role institutions play in addressing tolerance and peace has shifted from what was once perceived as reactive and temporary to what many now recognize as proactive and necessary. As such, institutional approaches that have typically implemented initiatives targeting specific areas must now employ a more comprehensive approach that encompasses multiple areas, which extends beyond compliance, status quo, and business as usual. As the need for tolerance and peace continues to grow — individually, institutionally, systematically, and structurally — the practice of equity, inclusion, and diversity allows for institutions to truly see the integral value of a more expansive approach that can focus on three areas: leadership, ownership, and partnership.

3.2 Leadership

Why is staying within the silos of "status quo" so natural and preferable? A sociological response could be that what is known breeds a certain level of security and comfort, but a reply from a pedagogical perspective would question whether security and comfort should, in fact, be goals at all. This has been the question at the core of the pursuit for tolerance and peace. With either response, an important understanding is the dual nature of the endeavor: institutional as well as individual institution being the context, individual(s) being the content. Thus, the pursuit of tolerance and peace must address leadership dynamics, if it is to be successful.

According to Ronald Heifetz and Marty Linsky, leadership would be a safe undertaking if institutions faced problems for which they already have the solutions.[1] This is a critical lens through which to view the pursuit, and practice, of tolerance and peace as it distinguishes *technical* challenges — those which people have the necessary know-how and procedures to tackle from *adaptive* challenges and those that require experiments, new discoveries, and adjustments from numerous places within the institution.[2] As adaptive challenges present themselves, the tendency is for members of an institution to look to an expert to provide a technical solution: "Tell me/us what to do." This approach allows institutions and individuals to avoid the dangers, either intentional or unintentional, of risk and vulnerability by treating adaptive challenges as technical. Which is why, Heifetz and Linsky conclude, management is more prevalent than leadership.[3]

To define leadership as an activity that addresses adaptive challenges considers not only the values that a goal represents "but also the goal's ability to mobilize people to face, rather than avoid, tough realities and conflicts."[4] The most difficult and valuable task of leadership in the area of tolerance and peace may be advancing goals and articulating strategies that promote adaptive solutions — undertaking the iterative process of examining where an institution is, how it arrived to that point, and what it needs to do to move.

---

[1] Heifetz and Linsky, *Leadership on the Line: Staying Alive through the Dangers of Leading*, 13.

[2] Ibid.

[3] Heifetz and Linsky, *Leadership on the Line: Staying Alive through the Dangers of Leading*, 14.

[4] Heifetz, *Leadership without Easy Answers*, 23.

In other words, a big-picture perspective, fueled by the need for change and immersed in constant action.[5]

The assertions of Heifetz and Linsky suggest that, regarding tolerance and peace, leaders cannot simply recognize the challenges facing an institution but must be careful to understand their historical *and* structural nature while interpreting them in adaptive terms. If, as Barbara Crosby and John Bryson argue, in order to coordinate action and make headway on resolving a complex institutional problem, those involved need to be aware of the whole problem system and recognize that it has to undergo significant change;[6] engaging the broad scope of systems and structures at both the macro and micro levels is essential for leadership. Challenges facing institutions require a process that addresses the various dimensions collectively. The pursuit of tolerance and peace, in an equitable fashion, would have to include careful attention to the role of culture, commitment, and communication. Yet, throughout this process, a clear focus on leadership, perhaps even shared leadership, must be employed to avoid institutional pitfalls of isolation, perpetuation, and stagnation. John P. Kotter explains that "needed change can still stall because of inwardly focused cultures, paralyzing bureaucracy, parochial politics, a low level of trust, lack of teamwork, arrogant attitudes... and the general human fear of the unknown."[7]

Given the climate of polarization, it would be prudent for leaders to imagine a different manner by which change occurs in institutions. Traditional strategic planning processes, in which an identified "upper" management team creates a plan with a set of "SMART" goals — specific, measurable, achievable, realistic, time-framed — and then systematically disseminates it into the system for implementation, tend to incorporate linear methods to obtain complex, dynamic outcomes. Since, according to Kotter, attempting to create major change with simple, linear, analytical processes almost always fails,[8] *situational management* must be replaced with *shared leadership*. For certain institutions, such as higher education, it is very difficult, arguably impossible, to enforce change given certain realities such as tenure, faculty approval, and board consent. As a result, leadership that engages across silos embodies a transformational mindset that mobilizes versus manages — a

---

[5]Heifetz and Linsky, *Leadership on the Line: Staying Alive through the Dangers of Leading*, 53.

[6]Crosby and Bryson, *Leadership for the Common Good: Tackling Public Problems in a Shared-Power World*, 9.

[7]Kotter, *Leading Change*, 20.

[8]Ibid., 25.

traditional approach that will be unsuccessful because it inadequately engages the breadth and depth of the institution.

John Kotter defines management as a set of processes that can keep a complicated system running smoothly, whereas leadership is a set of processes which creates or adapts institutions to significantly changing circumstances.[9] Unfortunately, a management mindset has been institutionalized, resulting in a culture that discourages leaders from learning *how* to lead. Ironically, this institutionalization is a direct result of past successes — the repetitious pattern of "doing what has always been done." Kotter diagnoses the syndrome as follows:

> Success creates some degree of market dominance, which in turn produces growth. After a while, keeping the ever-larger organization under control becomes the primary challenge. So attention turns inward, and managerial competencies are nurtured. With a strong emphasis on management but not leadership, bureaucracy and an inward focus take over. But with continued success, the result mostly of market dominance, the problem often goes unaddressed and an unhealthy arrogance begins to evolve. All of these characteristics then make any transformation effort much more difficult.[10]

The combination of both institutions that resist change and leaders who have not been taught how to create change is lethal, particularly because sources of complacency, status quo, and business as usual are rarely adequately addressed. Urgency and change are not issues for those that are comfortable with, and who have been asked to simply maintain, a current system of policies, processes, and practices.[11]

In addition to practicing situational management versus shared leadership, another common error made by institutions is to pursue change in a non-integrative manner. Crosby and Bryson distinguish two types of organizations: in-charge, or hierarchical, and networked, or shared (Figure 3.1). The in-charge organization has at its apex an individual or small group that establishes organizational direction, determines guiding policies, and transmits directives downward. Embedded in this type of organization is the assumption that the organization engages in highly rational, expert-based planning and

---

[9] Kotter, *Leading Change*, 25.
[10] Ibid., 27.
[11] Ibid., 29.

informed decision-making at the highest level. Highlighting the inadequacy of this structure, Crosby and Bryson see a need for a networked approach that includes a variety of cross-stakeholder engagement and inclusivity as a better, more beneficial model to influence change.[12] As they point out:

> Change advocates have to engage in political, issue-oriented, and therefore messy planning and decision making, in which shared goals and mission are being developed as the process moves along. New networks must be created, old ones co-opted or neutralized. These networks range from the highly informal, in which the main activity is information sharing, to more organized shared-power arrangements.[13]

For the networked model to be effective, two premises must be accepted. First, a certain loss of autonomy will be experienced. Within a hierarchical model, lower levels may possess knowledge but lack the trust in, and relationship with, upper levels. As a result, that knowledge may not get shared. Here, an approach that Geoffrey Vickers calls "acts of appreciation" becomes a useful lens because appreciation merges judgment of what is *real* with judgment of what is *valuable*. Identifying problems involves new appreciation of how something works, what is wrong with it, and how it might become better — from multiple perspectives. This appreciation subsequently shapes the way a problem is defined, the solutions considered, and the experiences of those impacted.[14]

Second, an understanding of culture is pivotal. Edgar Schein distinguishes three levels of organizational culture: artifacts, which are visible organizational structures and processes; espoused beliefs and values, which are strategies, goals, and philosophies; and underlying assumptions, which are unconscious, taken-for-granted beliefs, perceptions, thoughts, and feelings.[15] Institutional culture is inextricably linked to historical and current realities. Not accepting this inter-relationship will undermine the efforts of any network with which it becomes involved. As Schein points out:

> "The most central issue for leaders, therefore, is how to get at the deeper levels of a culture, how to assess the functionality of the

---

[12] Crosby and Bryson, *Leadership for the Common Good*, 5.
[13] Ibid., 9.
[14] Crosby and Bryson, *Leadership for the Common Good*, 15.
[15] Schein, *Organizational Culture and Leadership*, 26.

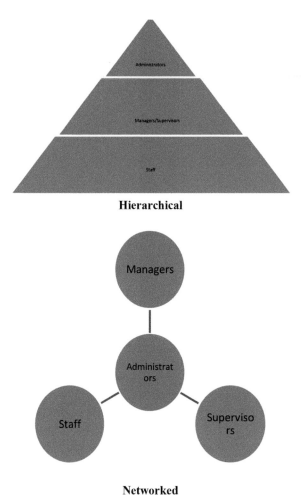

Hierarchical

Networked

Figure 3.1 Hierarchal vs. Networked.

assumptions made at that level, and how to deal with the anxiety that is unleashed when those levels are challenged."[16]

Schein defines culture as "a pattern of shared basic assumptions that was learned by a group as it solved its problems of external adaptation and internal integration, that has worked well enough to be considered valid and, therefore, to be taught to new members as the correct way to perceive, think,

---

[16] Ibid., 37.

and feel in relation to those problems."[17] Given this definition, one can see not just the historical connection between institutional culture and leadership but also the problem posed by a networked model: any challenging or questioning of basic, underlying assumptions will release anxiety and defensiveness.[18]

For purposes here, the use of "culture" builds on Raymond Williams's notion of culture "as the study of relationships between elements in a whole way of life."[19] From a leadership recognition standpoint, I assert a slight modification: *Systems do as designed, individuals do as allowed.* The pursuit of tolerance and peace requires a leader to shepherd and enhance ownership. While managers seek to control ownership, leaders must work to inspire ownership. Managers approach ownership from a hierarchical perspective; leaders, on the other hand, are more global in their approach to ownership.

3.3 Ownership

To effectively promote and practice tolerance and peace, an institution needs to establish some level of ownership in terms of where it is, how it arrived there, and what needs to be done. In most instances, policies, processes, and practices have a direct correlation to outcomes. A lack of ownership, or the practice of deflection, not only can manifest itself on an individual level but can also be fostered by an institution that tolerates conditions that contribute to intolerance and oppression. Institutional habits, left unchecked and uncorrected, can encourage a lack of ownership and undermine trust and transparency, resulting in unclear priorities, silo mentalities, and habitual conflict avoidance that invariably take away from addressing key issues — individually and institutionally — associated with, and necessary for, a culture of tolerance and peace. To combat this, ownership should involve an embracing and unpacking of institutional history, considering its effects at various levels, as well as its impact across the institution, leaving no structure, function, or area unexplored.

Power and privilege dynamics are constant institutional realities that can either help or hinder attempts at ownership. To make strides toward establishing accountability, institutions must begin with the frank acknowledgement that there are embedded causes of persistent, patterned orders of inequity — specifically, who has access and to what extent. This entails barriers and

---

[17]Ibid., 17.
[18]Ibid., 32.
[19]Williams, *The Long Revolution*, 63.

constraints that are more burdensome for those with the least amount of power and least access, leading to "meaning-systems" that, "while originally only ideas, gain force as they are reproduced in the material conditions of society."[20] The power and privilege dynamics within institutions stem from the acceptance of social mindsets that result in conditions becoming a part of, and reinforcement for, contingent applications and meanings — directly resulting in distrust; limited, if any, inclusion; and lack of communication.

Because culture is a critical component, it is essentially the construct that establishes values, practices, and, most importantly, sanctions that mark the institutional way of life. In the final analysis, it comprises what people do, how they go about doing it, and the impact throughout.[21] While the modern use of the term *culture* obscures the original, dynamic, and creative meaning of "tending, harvesting, or cultivating," retaining this active sense alerts us to the fact that culture is not some inert abstract reality but is always in process, in that it is always affecting and always being actively produced. Specific historical context may inform culture, but different content influences it. Consequently, culture is not a monolithic stationary entity that should be rejected, accommodated, or even transformed but rather is an existential reality that exists in a critical, discriminating, and constructive manner.[22]

So, what is the implication for tolerance and peace? Mirroring a world-view predicated on inequitable social structures directly places institutions in patterns of domination and subordination, possibly oppression. For these patterns to be purged, the behavior must be identified and addressed as a critical response to the need to achieve outcomes — healthy, positive, or otherwise — for all persons, groups, and stakeholders of the institution. Under these circumstances, ownership must be understood and approached in terms of *challenge*. Similarly, as an equity-building component, culture needs to be rediscovered as a cultivating process that creates standards, models expectations, and addresses behavior.

Another key component to ownership is the development of metrics that intentionally and progressively move the institution in a critical way. The management adage that "what gets measured gets valued" is particularly relevant. Because the biases that perpetuate intolerance are largely unconscious and reflexive, shifting an institution's emphasis from "fit" to "need" requires more than the "good intentions of well-meaning people." Without clear and

---

[20] López, *White by Law: The Legal Construction of Race*, 10.

[21] Tanner, *Theories of Culture: A New Agenda for Theology*, 27.

[22] Ibid.

robust measures to track equity efforts and outcomes, a tendency to revert to habitual, ingrained thinking, and behavior patterns restricts innovative investment and measurement. Metrics help institutions avoid the types of traps identified by Banaji and Greenwald as "mindbugs" — ingrained habits of thought and approach that lead to errors in perception, remembrance, reasoning, and decision-making.[23]

Strategically used, metrics can prioritize initiatives, establish targets, and, most importantly, evaluate impact. This institutional aligning of metrics can then serve as evidence of commitment while serving a cultural purpose as well, prioritizing engagement with and exposure to *difference*. These two elements, as well as the factors associated with them, must begin with deep dives into the past and present institutional realities — policies, processes, and practices that impact the creation and sustaining of culture. This unpacking of historical and current realities is a critical step in revealing the developmental aspects and effects relative to current climate, comfort, and achievement.

It is crucial that all levels throughout the institution model the importance of tolerance and peace. From senior leadership to front-line staff, the entire community must take ownership and be held accountable for application and advancement. The leadership team, by its actions or lack thereof, can signal importance through active involvement in the development of equity goals, designation in the strategic plan, and articulation in mission. Staff must work to ensure that department and team interactions model the institution's emphasis on tolerance and peace. Too often, institutional activities lean toward preferred exclusion — for the sake of "safe spaces," "avoiding conflict," and support of individualism — rather than intentional inclusion that can contribute to learning, growth, and development.

Owning the responsibility of creating a tolerant and peaceful-minded professional corps is also essential. Diversity requires more than numbers, and inclusion demands more than a superficial seat at the decision-making table. All aspects of the reward system must be continually reviewed and renewed from within an equity framework. For example, in higher education, the work that many underrepresented faculty and staff do with under-represented students must be valued in professional annual assessment systems such as

---

[23]Mahzarin R. Banaji and Anthony G. Greenwald, *Blindspot: Hidden Biases of Good People*, 4. While the authors do not link the term to institutional bias explicitly, the broader connection between individual, institutional, and larger society is made clear — "understanding how mindbugs erode the coastline of rational thought, and ultimately the very possibility of a just and productive society, requires understanding the mindbugs that are at the root of the disparity between our inner minds and outward actions." 20.

tenure and promotion; assessment methods that identify creating a tolerant and peaceful campus must be in staff evaluations; and the demographic makeup of teams, areas, and departments should be a criteria for leadership evaluation. Development opportunities could include, but not be limited to, education courses, professional workshops, and action research projects — all of which not only enhance individual competency, but institutional capacity as well.

3.4 Partnerships

Building a culture of tolerance and peace is both institutional as well as individual — a "both/and" versus an "either/or" construct. The very nature that it involves both capacity and competency is problematic because it flies in the face of equality — whose historical understanding and practice implies that once equal rights were achieved, individual ills and conditions of under-represented groups would be remedied. This ignores that a certain majority holds the structural bloodlines "in society to infuse their racial prejudice into the laws, policies, practices, and norms of society."[24] It is ironic that such an understanding and practice would be so misapplied, given that institutions have historically dealt with the problematic in ways that have recognized the underlying need for, and practice of, equity (e.g., the United States G.I. Bill), thereby making social transformation possible and standing as a structural principle in democratic idealism — *access*.[25]

Partnership is a powerful means to create access. It becomes a way of thinking that transforms silo mindsets into innovative pathways. Partnership, when done well, invites commitment, eliminates competitiveness, and encourages a sense of belonging. Internally, partnerships can help diagnose problems more comprehensively and clearly; externally, they may help identify sources that can help provide better solutions. The view that education, healthcare, or any other "system" is solely the responsibility of institutions in those systems is a fledgling concept. Each sector, whether by choice or force, has taken a decentralized approach in which partnerships have become not just beneficial but essential. Simply put, partnerships work to define problems more broadly, expand strategic thinking, and explore collective solutions.

---

[24]Robin DiAngelo, *White Fragility: Why It's So Hard for White People to Talk About Racism*, 22.

[25]See J. Kēhaulani Kauanui, "A Structure, Not an Event: Settler Colonialism and Enduring Indigeneity," in *Lateral: Journal of the Cultural Studies Association* 5, no. 1, 2016.

Whether referred to as "community engagement," "civic engagement," or "campus partnerships," collaborative constructs can assist with institutional capacity. Partnerships can involve a variety of areas, levels, and entities that can help with a wide range of issues including lack of recognition, resources, and ability to respond. Many of the partnerships necessary to create successful strategies for tolerance and peace will involve building pathways of imagination and innovation, inside and outside the institution. Approaching these as authentic relationship-building, opportunities can be an integral step in building trust, removing misconceptions, and contributing to the realization that the need for relationships may not just be prudent but also transformational.[26]

Martin Luther King called the art of alliances complex and intricate.[27] It can be argued that his assertion was accurate because building alliances is much more detailed than putting exciting combinations and ideas on paper. It involves an acknowledgement of self and common interests, validation of individual and group identity, in addition to affirmation of isolated and shared resources. If, as King argued, we employ the principle of selectivity along these lines, we will find millions of allies who, in serving themselves, also support the various institutions that house them, "and on such sound foundations unity and mutual trust and tangible accomplishment will flourish."[28]

Another aspect that is advantageous to explore is *who*, or *what*, has power — defined as the ability to construct, control, coerce, and change. When Cornel West speaks of perpetrators of free-market fundamentalism and authoritarianism, he defines them as "plutocratic leaders, corporate elites, elected officials, [and] arrogant authoritarians."[29] In other words, they are those in socially constructed positions who have the ability and authority, based on access, to designate the parameters of association. In civic institutions, such persons would be presidents, executive directors, or other "gate-keepers" who can make significant decisions with profound institutional impact. This is important because true authentic engagement begins with trust, transparency, and robust relationships.[30]

---

[26] See Lee G. Bolman and Terrance E. Deal, *Reframing Organizations: Artistry, Choice, and Leadership.*

[27] Washington, *A Testament of Hope*, 309.

[28] Ibid., 310.

[29] West, *Democracy Matters*, 21–23.

[30] Ibid., 28.

As they studied successful change efforts, Crosby and Bryson "realized that organizations had to find a way to tap each other's resources (broadly conceived) in order to work effectively on public problems. That is, they had to engage in sharing activities, which vary in level of commitment and loss of autonomy."[31] This brings to bear a critical point: most leaders and institutions are either unwilling or uncomfortable to forfeit autonomy and power. This is especially true where hierarchy is the tradition. For institutions to advance, the philosophical approach must change to visualize what can be accomplished by a shared-power structure that otherwise without, renders institutions less informed, responsive, and resourceful. Shared-power arrangements may be most useful in creating a climate where those with little to less institutional *authority* feel a sense of *creative deviance* that enables them to step away from providing answers that soothe and readily raise questions that disturb.[32]

Distinguishing silos can be a challenge to creating partnerships. There are several sources of the silo mentality that can affect institutional culture: areas of expertise, learned behavior, and unwillingness to think broadly across the institution, just to name a few. Structure and culture of the institution can also foster a silo mentality. If institutions do not establish cross-functional meetings, training, and development sessions, or even impact-evaluation discussions related to policies that bring people from different areas, departments, and levels together, individuals will remain in their "caves of comfort." It is imperative that institutions create and promote a culture that prioritizes sharing, collaborating, and "outside the box" thinking.

3.5 Conclusion

I have offered a networked approach to tolerance and peace that is made more effective through integration with equity, diversity, and inclusion. It is my belief that specific dimensions of this approach are critical and anticipate that this perspective will assist institutions in recognizing crucial areas and aspects to address institutional capacity as well as individual competency. In one study, more than 80% of companies identified leading change as one of the top five core leadership competencies for the future. More importantly, 85%

---

[31] Crosby and Bryson, *Leadership for the Common Good: Tackling Public Problems in a Shared-Power World*, 17.
[32] Ronald A. Heifetz, *Leadership Without Easy Answers*, 188.

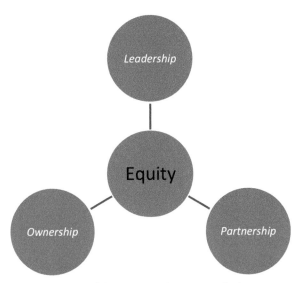

Figure 3.2   Hierarchal vs. Networked.

felt that this competency was not as strong as was needed.[33] Make no mistake, tolerance and peace is about change, on an institutional and individual level. A networked model that gives attention to leadership, ownership, and partnership is critical. As we engage much broader ways of thinking about tolerance and peace, institutions need to expand their approach to encompass the internal and external dynamics associated with equity, diversity, and inclusion. Approaching tolerance and peace in this manner allows strategies to pull together the use of competency, capacity, and community to formulate institutional plans where constituents and stakeholders are equipped, accountable, and connected.

---

[33] See J. Stewart Black and Hal B. Gregersen, *It Starts with One: Changing Individuals Changes Institutions.*

4

An Identity-Based Conceptual Framework for the Assessment of Tolerance in Education Curricula

Karina V. Korostelina

George Mason University

Abstract

Tolerance is a complex, multifaceted phenomenon that could be analyzed on three levels: individual, intergroup relations, and society. A social identity-based approach helps to understand major foundations of tolerance through the analysis of the dynamics of identity-based conflicts. This chapter describes the process of building tolerance in education as a continuum that progresses from *incitement to violence and hatred* (intense dislike and hatred of an outgroup, justified willingness to fight with or harm outgroup members) to *prejudice* (an unjustified or incorrect negative attitude toward members of an outgroup based on the membership of the ingroup) to *tolerance* (an acceptance of the Other — a fair, objective, and permissive attitude toward those whose opinions, beliefs, practices, racial or ethnic origins, etc., differ from one's own) and then to *mutual understanding* (including critical analysis of conflict, empathy, compassion, and willingness to cooperate). The chapter proposes an assessment framework based on 14 indicators, including: salience and forms of identity, metacontrast, prototypes, favorable comparison, projection, social boundary, relative deprivation, collective axiology, etc. This assessment framework can serve as a foundation for the development of textbooks and teaching manuals as well as an overall assessment of curricula and education policies.

4.1 An Identity-Based Conceptual Framework for the Assessment of Tolerance in Education Curricula

The concept of a "culture of peace" was first formulated by the International Congress on Peace in the Minds of Men, held in Cote d'Ivoire in 1989. In 1995, the 28th General Conference of UNESCO introduced the Medium-Term Strategy for 1996–2001 (28 C/4) centered around this concept: "To counter the culture of war, let us build a culture of peace, that is to say a culture of social interaction, based on the principles of freedom, justice and democracy, tolerance and solidarity, and respect for all human rights; a culture that rejects violence and, instead, seeks a solution to problems through dialogue and negotiation; a culture of prevention that endeavors to detect the sources of conflicts at their very roots, so as to deal with them more effectively and, as far as practicable, to avoid them" (UNESCO, 1995). In 1997, the 52nd session of the United Nations General Assembly discussed the specific topic — "Towards a Culture of Peace" — and proclaimed the year of 2000 as the International Year for the Culture of Peace. The decade of 2001–2010 was defined as the "International Decade for a Culture of Peace and Non-Violence for the Children of the World" by the 53rd session of the United Nations General Assembly that adopted the Resolution A/53/25 based on the proposal of the group of Nobel Peace Prize laureates. Over 75 million people around the globe (more than 1% of the world's population) signed the common pledge drafted by the Nobel laureates to promote the universal principles of a culture of peace and non-violence in daily life.

Since then, multiple documents and resolutions of UNESCO emphasize the importance of a culture based on peace and tolerance. The UNESCO documents discuss that a culture of peace does not rest on a passive form of tolerance or abstract pacifism; instead, it creates clear paths to combat injustice, inequality, and oppression. The culture of peace functions as a "moral code in action" that requires profound obligation to create a just and peaceful world with values of human dignity and inclusion. The culture of peace is inconsistent with poverty, discrimination, and inequality and requires equal education, just distribution of wealth and knowledge, and development of democracy. Thus, the formation of a culture of peace and tolerance includes both the prevention of direct and structural violence as well as actively working toward creating a more just and equal world, as formulated in Galtung's (1969) concepts of negative and positive peace. While negative peace requires fighting against the culture of war, positive peace promotes a culture of tolerance, equality, and inclusion (Adams, 2002).

These positive and negative components of a culture of peace are evident in many societies, revealing the complex interplay of tolerance and peace; there is "persistence of social images of life at peace, the ineradicable longing for that peace, and the numbers of social movements working for a more just and peaceful world" (Boulding, 2006b). In their everyday life, people do not only try to avoid conflicts and violence but also negotiate differences and build resilient communities, creating international solidarity and unity in fighting injustices and equalities (Cromwell and Vogele, 2009).

The international community also formulated ways of building this culture of peace. The first International Forum on the Culture of Peace was organized in San Salvador (El Salvador) in 1994. Analyzing the formation of a culture of peace across the globe, the participants formulated three main means that are essential for this development: education, democratization, and participation. The educational component was further advanced at the 44th session of the International Conference on Education in 1994. Education (especially education for the peaceful resolution of conflict) is among UNESCO's eight areas of building of peace culture. This dimension of nurturance that includes tolerance, education, and equality was stressed as a critical aspect of a culture of peace (De Rivera 2004b).

One of the most important ways of achieving peace culture, according to these documents, is the development of a global identity that involves both local identities and a global solidarity against common threats to the planet. "The culture of peace may thus be defined as all the values, attitudes and forms of behavior, ways of life and of acting that reflect, and are inspired by, respect for life and for human beings and their dignity and rights, the rejection of violence, including terrorism in all its forms, and commitment to the principles of freedom, justice, solidarity, tolerance and understanding among peoples and between groups and individuals" (UNESCO, 1995). Boulding (2000b) also stresses the importance of understanding identity balance — between the need for autonomy and the need for relatedness — for both promoting tolerance and mutual understanding among people. She sees a peace culture as "a mosaic of identities, attitudes, values, beliefs... that lead people... to deal creatively with their differences and share their resources" (Boulding, 2000a, p. 196).

An identity-based approach to forming a culture of peace and tolerance through education emphasizes the reframing narratives of intergroup relations, the redefinition of conflict-based discourses, and the rehumanization of former enemies. This approach includes both negative and positive aspects of peace, discussed above, as it not only alters the negative representations of the Other but also challenges patterns of exclusion and inclusion, discrimination, and inequality based on belonging to particular social categories of ethnicity, religion, gender, and nation. "The culture of peace encourages peaceful interaction that refrains from violence and settles conflicts by improving positive relationships between the parties involved in various sectors of human life and activity: education, politics, economics, and daily routines. This culture, while acknowledging the differences that exist amongst humans and human groups, values such diversity as a source of richness and strength to the global community" (Korostelina, 2012, p. 6). An identity-based approach to tolerance encourages the celebration of diversity and mutuality, developing peaceful and just communities around the globe.

This approach explains how tolerance can be developed through education across ethnic, religious, and national lines by promoting the growth of new cultural forms out of old ones (Glaser and Strauss, 1967). First, education is a powerful vehicle for changing norms of exclusion and difference and developing new norms of inclusion and mutuality (Richards & Swanger, 2009). Second, an identity-based approach can help alter the meaning of intergroup relations, challenging the narratives of each side of the conflict and denigration of the Other, as well as in increasing one's own group's responsibility for the others' suffering (Salomon & Clairins, 2009). Third, it also promotes reflection on power, dominance, and categorical inequality in the creation of the culture of peace.

The chapter describes the process of building tolerance in education as a continuum that progresses from *incitement to violence and hatred* (intense dislike and hatred of an outgroup, justified willingness to fight with or harm outgroup members) to *prejudice* (an unjustified or incorrect negative attitude toward members of an outgroup based on the membership of the ingroup) to *tolerance* (an acceptance of the Other — a fair, objective, and permissive attitude toward those whose opinions, beliefs, practices, racial or ethnic origins, etc., differ from one's own) and then to *mutual understanding* (including critical analysis of conflict, empathy, compassion, and willingness to cooperate). The chapter proposes an assessment framework based on 14 indicators, including salience and forms of identity, metacontrast, prototypes, favorable comparison, projection, social boundary, relative deprivation, collective axiology, etc. This assessment framework can serve as a foundation for the development of textbooks and teaching manuals as well as overall assessment of curricula and education policies.

4.2 Indicators

Social Categorization:

Social categorization — the perception of people through their membership in social groups — defines how a person and others see her or his position in a society, impacts a person's self-image (Abrams & Hogg, 1988), helps a person make sense of the world (Reynolds Turner & Haslam, 2000), reduces ambiguity and uncertainty (Hogg, 2007), and leads a person to behave in ways that are consistent with the group (Hogg & Haines, 1996; Turner, 1975). Salience is the most important characteristic of identity that can vary on a continuum from strong to weak, influencing a person's attachment to the group as well as their loyalty and behavior to the group (Berry et al., 1989; Brewer, 1991; Brewer, 2001; Tajfel & Turner, 1979). Persons with a high salience of ethnic identity have a strong connection to other members of the group, positive feelings about the group, and a commitment to its values and goals (Phinney, 1991). Identity that remains salient for a long period becomes a central identity affecting a person's behavior.

Salient identity has significant impacts on how a person responds to different situations and is strongly linked to negative attitudes and violent behavior. Most studies on social identity provide evidence of a relationship between the salience of identity and attitudes toward outgroups. Other research results confirm the role of salient identity in shaping political attitudes and behavior (Conover, 1988; Miller, et al., 1981) and reveal strong correlations between salient identity and outgroup hostility (Branscombe & Wann, 1994; Grant & Brown, 1995). Any threat to beliefs and positions of the group reinforces the salience of social identity and can lead to collective actions (Brewer, 2007; Ting-Toomey et al., 2000). Salience of identity is constructed in education, mass media, and political discourse in multiple ways, including 1) stress on the importance of a particular (ethnic, national, or religious) identity, 2) multiple mentions of particular identity in comparison to others, and 3) references to the importance of loyalty to particular groups. The salience of identity and acceptance of an ingroup's norms impact intergroup forgiveness and reconciliation (Noor, Brown, Gonzalez, Manzi & Lewis, 2008; Wohl & Branscombe, 2005). For example, salient national identity has been associated with the support for strong responses to terrorist attacks and less concerns for human rights in the United States (Fischer, Greitemeyer & Kastenmuller, 2007), Europe (Strabac & Listhaug, 2008), the United Kingdom (Brighton, 2007), and Australia (Musgrove & McGarty, 2008; Strelan & Lawani, 2010). Thus, the emphasis on salience of a particular identity reduces the likeliness for reconciliation while the increased salience of common identity positively contributes to the reconciliation process (Staub, Pearlman & Hagengimana, 2005).

Social categorization theory (Tajfel & Turner, 1979) stresses the critical role of *group prototypes* in defining the meaning of group membership and understanding the norms, values, and behavior of group members. A prototype is a particular person who represents the most important beliefs and values of a group by serving as an emotional function and increasing cohesiveness within the group. A prototype focuses on the similarities within an ingroup, which strengthens social identity — differences found from those of the prototype are perceived as less attractive and even unfavorable. A leader as a prototype could promote conflict intentions, establish enemy images and threat narratives, and motivate group members to continue the fight. Alternatively, a prototype can promote values and ideas of peace and forgiveness, increasing groups' participation in reconciliation practices. Thus, a promotion of specific ingroup prototypes can impact the reconciliation process. The impediments can come from the employment of main historical figures who are warriors or protectors of faith, people who sacrificed their life, or individuals who led people to fight with the other group as the enemy (Korostelina, 2013). Usually, violent actions of these prototypes (e.g., suicide attacks, fighting, etc.) are positively evaluated. To support reconciliation processes, it is important to endorse prototypes that are peaceful, forgiving people who promote tolerance and coexistence.

Social categorization theory also emphasizes the role of *metacontrast* — the perception that differences within an ingroup are smaller than those between ingroup and outgroup — in intergroup relations. A high level of metacontrast enhances intergroup differences, reduces understanding, empathy, and compassion. For example, some religious doctrines strive to establish intrareligious hegemony by maximizing the contrast with the dominant outgroup; Christianity or Judaism, for example, maximizes intergroup difference to produce extreme, maximally counterintuitive concepts (Nicholson, 2014). The greater the perceived difference in the typical characteristics of the ingroup and the outgroup, the greater the predisposition to hostility (Oakes, 1987; Turner et al., 1994). Metacontrast can be boosted in ingroup narratives through the description of all ingroup members as similar to each other, having the same destiny, goals, and aspirations as well as through the stress on differences with the outgroup and its supporters. Such employment of metacontrast can negatively impact reconciliation processes by stressing differences between groups and fostering ingroup homogeneity.

4.3 Negative Attitudes Toward the Other

The theory of social identity suggests that people have an essential need to acquire high social status and a positive identity through membership in socially prestigious groups. This search for positive self-esteem is the basis for the formation of negative attitudes toward outgroups (Brown, 2000; Huddy & Virtanen 1995; Jackson et al., 1996; Tajfel & Turner, 1979; Taylor et al., 1987; Wright, Taylor & Moghaddam, 1990). This basic need leads to the *favorability comparison* — the tendency to evaluate outgroups negatively in contrast to the ingroup that results in the formation of positive stereotypes related to ingroup members and negative stereotypes related to outgroup members. Thus, a favorable comparison develops perceptions of outgroups as a "second sort" of people, leading to prejudices and blatant discrimination. The need for favorable comparison is even more important if groups have a low economic and social status, have a minority position in society, or lack the opportunity to promote, devellop, or revive their culture. Reconciliation processes can be highly impacted by these favorable comparison processes, especially if ingroup narratives represent the ingroup as superior to the outgroup, including its culture, history, religion, values, and traditions. To combat such comparison, ingroup narratives could concentrate on internal locus of self-esteem (Korostelina, 2007) that can be archived through 1) the emphasis on rich cultures, famous artists, writers, scientists, and engineers, 2) previous exceptional positions or roles throughout history, and 3) their uniqueness or exceptionality. If ingroup members are proud of their identity and have a high sense of confidence, they have a lesser tendency to use a favorable comparison between their groups and outgroups.

Global attribution (Allport, 1954; Pettigrew, 1979) is described as a tendency for people to over-emphasize dispositional, or character-based, explanations for behaviors observed in outgroups while over-emphasizing the role and power of situational influences on ingroup behavior. Pettigrew (1979) described a tendency of ingroup members to make internal (dispositional) attributions for successes of the ingroup and external (situational) attributions for the ingroup's failures while making internal (dispositional) attributions for the outgroup's failures and external (situational) attributions for the outgroup's successes, leading to the fundamental attribution error. Hewstone (1989) reviewed many studies documenting the fundamental attribution error and found that this error leads to increased conflict between groups. Global attribution error provides justifications of aggressive ingroup actions as a response to the situation created by the intentions of the outgroup.

In ingroup narratives, actions of an outgroup are interpreted in terms of their harmful and aggressive motivation and goals, while actions of ingroup are interpreted in terms of response to the situation (often created by the outgroup). Such perception reduces success of reconciliation processes by placing the responsibility for aggression and violence on the outgroup and denying accountability of the ingroup.

Psychodynamic theory (Volkan, 1997; Volkan 2004) describes this process of the justification of ingroup actions by putting the blame on the outgroup as *projection*. As a person inclined to deny negative characteristics of herself or himself, groups also tend to project negative images into others (Volkan, 1998). More specifically, people tend to split off and externalize negative aspects of oneself — the characteristics they do not want to acknowledge or take responsibility for. Group identity is perceived as a "large canvas tent" that shields group members from external threat (Volkan, 1998, p. 27). According to the theory, as long as this tent remains robust and steady, the ingroup members are not conscious of its role and do not need to continuously ascertain or define their group identity. But when the tent becomes unstable or disturbed, ingroup members raise their collective concerns and work together "to shore it up" again (Volkan, 1998, p. 27). In these situations, ingroup members project their negative features on outgroup members. Projection externalizes and ascribes to outgroups the negative characteristics of the ingroup or explains ingroup actions by provocation of the outgroup. Thus, the ingroup can justify its aggressiveness by the need for defense provoked by the threatening actions of an outgroup. Or, the ingroup can validate its disloyalty by attributing treasons to the outgroup. The ingroup narratives can emphasize descriptions of ingroup actions as protective in response to outgroup aggression, rejections of peace and coexistence based on the description of the outgroup as liars, deceivers, and not trustworthy, and justifications of ingroup violent actions by describing outgroup as a provocateur. Similar to global attribution error, such narratives can impede reconciliation processes by denying ingroup responsibility and reducing willingness to collaborate with the vicious outgroup.

4.4 Forms of Identity

To understand the impact of social identity on the formation of a culture of tolerance, it is important to concentrate not only on salience but also on the meaning of group identities. One of the ways to explore identity meaning is to look at the specific forms that a particular identity can take, including: cultural, reflected, and mobilized (Korostelina, 2007). The *cultural form of identity* is rooted in the poetics of everyday life of a group, involving cuisine and diet; attires and costumes; typical daily routines; music, songs, and dance; rituals and habits; and festivals, holidays, and special ceremonies for festivities or grief. While beliefs, positions, and norms are essential for this form of identity, they are considered given and foundational and usually do not become a subject of reflection. Individuals live "within" their cultural identity, abiding by all ingroup norms and prescriptions but never question the values, aims, and intents of their ingroup, and assess relations between ingroup and other groups based on cultural differences and similarities. As cultural forms of social identity do not aid in deepening the understanding of the meaning of ingroup identity and connotation intergroup relations, any violations (even without any intention to infringe) of specific cultural rituals, norms or customs could be perceived as a threat to ingroup identity, inciting conflict intentions, and reducing the culture of tolerance. Such cultural forms can be formed in education curricula through 1) presentations of traditions, customs, and cultural holidays as rightful, 2) avoidance of discussions about the historic development of national identity, roots, and meaning of cultural traditions, and 3) negative representations of other cultures.

The *reflected form of identity* is associated with a deeper comprehension of the history of the ingroup and its relations with outgroups; it refers to the attentiveness of the social status and place of the ingroup in a society in addition to an understanding of its aims and perspectives. This identity form also concentrates on values and beliefs of the group with a deeper knowledge about its historic roots and an acknowledgement of the position of the group among other groups in a society. The reflected form has a strong prospective to become a foundation of the formation of a culture of tolerance, as it is rooted in an advanced comprehension of ingroup values and goals as well as an understanding and appreciation of the differences between groups. The reflected form also supports a deeper consideration for positions and activities of ingroup and outgroup members, analyzing intergroup relations from a more balanced point of view. Such reflected form can be developed in education curricula through 1) increasing awareness of the history, roots, and sources of the ingroup, its relationship to outgroups, and the current status, position, and perspectives of the ingroup; 2) an emphasis on understanding of common history and shared goals with outgroup; and 3) the presentation of the roots and meanings of regional cultural traditions and beliefs that can unify nations.

The *mobilized form of identity* is based on the view of ingroup identity from the standpoint of intergroup relations, concentrating on comparisons between groups' power, status, and problems in intergroup relations. The meaning of the ingroup arrives from the competitive assessment of the positions and goals of outgroups. While a mobilized form of identity also includes customs, values, and cultural characteristics, they are less essential than this intergroup comparison. Such ideologization of mobilized identity leads to the perception of competition, contradictions, and incompatibility of goals between the two groups (Korostelina, 2007). The meaning of mobilized identity centers around the need to increase the status or power of the ingroup, readiness to compete or fight against the outgroup that results in negative intergroup relations. Such mobilized forms can be formed in education curricula through 1) depicting of the aims, values, and ideas of a particular nation as the only possible or rightful way of thinking, 2) praising national leadership as the only ones capable to lead a nation, demanding faithfulness and submission to this leadership, 3) presenting of members of outgroups as adversaries, and urging students to unite against their continuous demands, and 4) stressing the intergroup interaction of a "we–they" opposition perspective.

4.5 Interrelations Between Groups

A *social boundary* is a crucial mechanism of the formation of social identities which defines not only the relationship between groups but also the meaning of the ingroup identity (Barth, 1981). The social boundary is a distinctive narrative about relations on both sides of the boundary and across the boundary that is formed as people cultivate and sustain relations within their groups as well as develop interrelations between the groups across this social boundary (Tilly, 2005). These narratives form the foundations for collective identities and define meaning of boundaries as sites common to two groups (Eyal, 2006; Thelen, 2002).

Social boundaries are contingent on a variety of contextual factors, including the cultural repertoires, customs, and dominant narratives in a particular group as well as on political movements or collective actions (Doevenspeck, 2011; Lamont, 2000; Somers, 1994; Swidler, 2001). The cultural and political elites also delineate social boundaries, outlining how encompassing, restricting, and accessible a particular social boundary should be (Horowitz, 1975), creating social order, defining and classifying relations between social groups (Tajfel and Turner, 1979). Social boundaries are formed through the creation of new narratives of difference or borrowing of different boundaries, encounters between previously distinct or competing groups, and shifting meanings of ingroup identities (McAdam et al., 2001). A metacontrast — a tendency to minimize the intragroup differences and maximize the intergroups dissimilarities, as discussed above — make social boundaries more impermeable (Tajfel & Turner, 1979).

Together with established and institutionalized categories as foundations for social boundaries (ethnic, national, religious, etc.), various conceptions, interpretative schemata, and cultural dimensions also can contribute to the development, maintenance, and contestation of differences between social groups (Lamont et al., 2015). Such *symbolic boundaries* establish essential distinctions, contesting and redefining the meaning of established social boundaries. In some cases, symbolic boundaries become so salient that they replace social boundaries (Lamont & Molnar, 2002). In these cases, symbolic boundaries become mechanisms for contentious politics, challenging or preserving existing power relations and patterns of exclusion and inclusion, and opportunity hoarding (Bourdieu, 1977; Gramsci & Lipset, 1959; Tilly, 2003; Tilly, 2006).

The education curricula can make a social boundary more impermeable, supporting social hierarchies, discrimination, and exploitation by 1) the removal of any history of positive relations, traces of interaction, and descriptions of shared living spaces from textbooks; 2) the denial or downplaying of similarities between groups and emphasis on differences as unsolvable and permanent; 3) the defining of the ingroup and outgroup as distinct groups with different histories, divergent core values, and paths of development; 4) the promotion of the dominance of the ingroup over the outgroup and denial of the cultural rights of the outgroup; and 5) the stress on the controversial and disputed aspects of history and the roots of conflicts, misunderstandings, and historical divides.

It can also challenge existing social and symbolic boundaries, making them more open or creating new social and symbolic boundaries that reduce inequality and exclusion and promote tolerance. It can be done through 1) a shift of perspective from ingroup histories to a common approach to history and emphasis on common tendencies and transnational processes; 2) the creation of an opportunity for ingroup members to understand the views of outgroups; 3) depiction of major concepts around society, politics, and international relations from both ingroup and outgroup perspectives; 4) promotion of a history of positive interrelations, common experiences, and collaborations; and 5) providing a balanced assessment of historical events based on a multiplicity of perspectives, comparison, and critical thinking.

4.6 Intergroup Competition

Intergroup competition is rooted in *relative deprivation* and horizontal inequalities when members of disadvantaged group perceive more discrimination and have more desire for social change (Gurr, 1970). Feelings of relative deprivation can arrive from the belief that the actual social or economic status of the ingroup is lower than the one expected by the group members (Davis, 1959; Runciman, 1966). It relates to the "perception of discrepancy between their value expectations and their value capabilities" (Gurr, 1970, p. 24). Temporal relative deprivation rests on the comparison between the past position of the group and its current situation, leading to longing for the "good old times" and myths of a Golden Age (Smith, 2011). To explain this relative loss, the ingroup usually blames outgroups and attributes them negative intentions.

Relative deprivation can also result from the comparison of the positions, resources, and power of ingroups and outgroups when members of the ingroup believe that they have less than they deserve in comparison to others (Runciman, 1966; Walker & Smith, 2001). This intergroup comparison leads to a strong belief that an ingroup is disadvantaged and unfairly treated, which invokes feelings of anger, bitterness, and entitlement (Pettigrew, 2015) with an increase in support for redistribution (Shin, 2018). Moreover, relative deprivation has a stronger effect on people's motivation than absolute deprivation (Smith, Pettigrew, Pippin & Bialosiewicz, 2012.)

Relative deprivation can reflect persistent inequalities, structural violence, and discrimination, thus promoting a change within existing economic and social policies. However, to influence the behavior of ingroup members, relative deprivation does not have to be real: just a perception of difference and disadvantage can provoke conflict intentions (Pettigrew & Tropp 2011). This tendency is often utilized by group leaders and political entrepreneurs to mobilize groups in fighting with outgroups or in supporting discriminatory policies toward them. History curricula can be used to emphasize relative deprivation and reduce tolerance. It can be done through 1) the portrayal of outgroups as having more rights and resources in comparison with the ingroup (fraternal deprivation); 2) stress on limitations of the socio-economic opportunities of the ingroup by outgroups (fraternal deprivation); 3) emphasis of unequal economic, cultural, or political positions of ingroups and outgroups (fraternal deprivation); 4) descriptions of the ingroup position as worsening over time (temporal deprivation); and 5) descriptions of the ingroup position as worse than it should be (deprivation as expectation).

4.7 Ingroup Victimization

Collective victories and defeats can be emphasized by ingroup leadership and influence a person's perceptions of intergroup relations. Volkan describes these perceptions as *chosen glories* (important, usually mythologized and idealized achievements that took place in the past) and *chosen traumas* (losses, defeats, humiliations—also mythologized—that are usually difficult to mourn). These chosen glories and traumas are usually rooted in actual events from the history of the group, functioning as "a shared mental representation of the event, which include realistic information, fantasized expectations, intense feelings, and defense against unacceptable thoughts" (Volkan, 1997). They can be passed from generation to generation as memories of un-mourned ancestors' trauma through the process of *transgenerational transmission*. The memories of ingroup tragedy are transmitted from one generation to the next: collective traumas that remain an unhealed wound emphasize that the ingroup had never achieved justice or retribution from the wrongs that befell their ancestors. Transgenerational transmission of trauma happens "when the mental representation becomes so burdensome that members of the group are unable to initiate or resolve the mourning of their losses or reverse their feelings of humiliation" (Volkan, 1997, p. 45). Chosen traumas are passed down to children and grandchildren in the hope that they may find a way to mourn and resolve persistent problems. New generations accept these memories and emotions as "psychological DNA" planted in their social identity. Thus, transgenerational trauma is trauma that is transferred from the first generation of trauma survivors to the second and further generations of offspring of the survivors via complex post-traumatic stress disorder mechanisms.

However, some of these events could have a small historical significance. They are chosen not because they were essential to the tradition or identity of the group or are passed through generations but because of the current conflicts and contradictions with outgroups. They provide "explanations" for poor economic conditions or minority status. Neisser (1967) describes mnemonic processes as an active construction that involves previous experiences, selection, distortion, and omission of information based on its importance for the reconstructed picture of the past. Collective remembering is a constant negotiation between past and present, "an active process of sensemaking through the time," (Warburg, 2010, p. 53), a mnemonic journey that encompasses never-ending reconstruction of the past and its meaning for present and future. Presentations of ingroup victories and defeats in history textbooks help students to unite around powerful ideas of group gains and losses and thus accept ingroup perceptions of intergroup relations.

The chosen traumas of the group can be promoted in educational curricula through an emphasis on the outgroup as extremely aggressive, vicious, and willing to destroy the ingroup through history; vivid descriptions of the aggressive acts of the outgroup in the past; presentation of the ingroup as an innocent victim of the aggressive, dominant outgroup; and a stress on the responsibility of new generations to remember traumas of their parents and to revenge them. Chosen glories are promoted through presentations of the ingroup as successful, with great achievements and glories, and praising the group for achievements. While it is important to provide an educational space to heal collective traumas, a strong emphasis on traumas reduces tolerance and acceptance of the Other. Highlighting ingroup glories can empower ingroup members and reduce the need for favorable comparison and negative perceptions of the Other.

4.8 Outgroup Threat

Realistic threats are threats to existence, (economic and political) power, and the (physical or material) well-being of the ingroup. Symbolic threats are connected to differences in values, morals, and standards between groups and depend on the perceived risks and challenges to the ingroup's worldview (Stephan et al., 2002). Studies show that both realistic (Bobo, 1999; Esses, Dovidio, Jackson & Armstrong, 2001; Ouillian, 1995) and symbolic (Esses, Haddock & Zanna, 1993; Sears & Henry, 2003; Stephan et al., 2002) threats increase the possibility that biases and prejudice will result in intolerance and discrimination. Other studies show that perceived threats to the ingroup link salient identity with negative attitudes toward the outgroup, ultimately leading to increased intergroup hostility (Johnson, Terry & Louis, 2005; Louis, Duck, Terry, Schuller & Lalonde, 2007).

If groups are in competition for meaningful resources, they will have a stronger feeling of outgroup threat, especially in situations when the conflicting groups have more to gain from succeeding. Numerous studies also show that outgroup threats create more intolerance among ingroup members toward the outgroup, justifying the conflict and the discriminative treatment of outgroup members. In situations of competition, proximity and contact increase feelings of threat and, thus, intolerance, rather than decreasing it (Brewer, 1972; Levine & Campbell, 1972; Sherif, 1966; Sherif & Sherif, 1953; Taylor & Moghaddam, 1994). Usually, the ingroup tends to perceive the outgroup as a threat in several contexts of intergroup relations such as the following: 1) unequal economic, cultural, or political positions of ethnic groups (Gellner, 1994); 2) minority status of ethnic groups (Brubaker, 1996); 3) memories of the former domination of the outgroup and attribution of the desire for revival (Gurr & Harff, 1994); 4) perceptions that groups have weaker or worse positions in comparison with the outgroup (Gurr, 1970); 5) limitations of the socio-economic opportunities of the ingroup by outgroups (Gellner, 1994); and 6) political extremism, violence, and nationalism of outgroups (Hagendorn, Linssen, Rotman & Tumanov, 1996).

Social groups are perceived not only as social units but also as organized entities with shared goals, intentions, and inspirations for the future. Thus, members of an ingroup usually see the outgroup not only through their culture, history, or behavioral features (stereotypes), but they also attribute goals to the outgroup (Blumer, 1958; Horowitz, 1985). These ascriptions of hostile and destructive goals lead the views on the outgroup as a threat to the well-being, status, and a very existence of the ingroup. As was discussed above, fundamental attribution error results in attribution of negative, rather than positive, attitudes and goals based on the tendency for people to over-emphasize dispositional, or personality-based, explanations for behaviors observed in others while under-emphasizing the role and power of situational influences on the same behavior (Heider, 1958; Jones & Harris, 1967; Ross, 1977). In the situation of perceived competition between groups, all actions of an outgroup are interpreted in terms of their harmful and aggressive motivation and goals, seen as a possible threat to an ingroup.

One form of the functioning of the outgroup threat is the security dilemma that can reshape social identities and provoke an identity conflict (Lake & Rothchild, 1998). The role of the security dilemma was analyzed on the level of international relations, including the Cold War (Jervis, 1978; Wheeler & Booth, 1992; Spear, 1996), as a source for ethnic conflicts (Posen, 1993; Snyder & Jervis, 1999), and the rise of nationalism (Ven Evra, 1999). Resulting from a zero-sum game perception, the understanding of any advance by an outgroup is seen as a loss for the ingroup. This interpretation rests on perceptions of uncertainty, mutual suspicion, and anxiety among ingroup members regarding the intentions of the outgroup toward them. While the intention to harm others may not be real, these fears increase suspicions and doubts, leading to violent actions: "It is one of the tragic implications of the security dilemma that mutual fear of what initially may never have existed may subsequently bring about exactly that which is feared most" (Herz, 1950, p.160). Competition between groups is deemed as a struggle for status, with the outgroup gain considered automatically as an ingroup defeat — leading to an increased perception of threat.

Another source of intolerance toward outgroups is a threat to intergroup boundaries: if social borders between the ingroup and outgroup are distorted and weakened, people have increased intentions to protect the distinctiveness of their group (Branscombe, 1999; Michael, Wohl, Nyla, Branscombe & McVicar, 2001). Concerns about the current position and future of the ingroup give rise to strong emotional responses in the form of collective angst (Wohl & Branscombe, 2009; Wohl, Branscombe & Klar, 2006). The more salient an ingroup identity of ingroup members is, the more important it is for them to ascertain and preserve a distinctive group identity and the stronger are the emotional effects of potential distinctiveness loss (Jetten, Spears & Postmaes, 2004). Similarity with outgroup members is perceived as a loss for the ingroup's essence. To differentiate the ingroup and the outgroup, people often react by emphasizing available dimensions of comparison (Brewer, 2001; Jetten & Spears, 2004). For example, to stress the distinction with an ethnic group, speaking a very similar language and preserving the uniqueness of one's own ethnic group and its political rights, people can over-emphasize some negative features of outgroup members and develop strong negative stereotypes.

Many studies show that a threat to positive group identity results in intolerance and discrimination against outgroups (Branscombe & Wann, 1994). A social identity threat arrives from the perceived decreased value of ingroup identity resulting from the recognition that the ingroup is discriminated against and is devalued by the outgroup. When group members, and especially those with salient ingroup identity, perceive threats to the ingroup, they tend to increase the relative positivity of their own group by derogating outgroups (Hornsey, 2008). This negative perception and evaluation of the outgroup can result from a perceived social identity threat, even if the outgroup is non-threatening and has low status (Cardinu & Reggiori, 2002).

In education curricula, outgroup threats can be presented through a variety of ways, including description of the ingroup as strongly threatened (physically and culturally) by an outgroup; presentation of the outgroup as having intentions to destroy the ingroup (kill all members and kill leadership), positioning the outgroup as denying identity and culture of the ingroup;

presentation of the outgroup as intending to dissolve identity and culture of the ingroup; description of the outgroup as willing to use all measures against the ingroup; and the emphasis on value differences with the outgroup seen as challenging the ingroup's worldview.

4.9 Collective Axiology

A collective axiology is a common moral and value system that offers moral guidance to ingroup members on how to perceive and treat members of ingroup and outgroups and how to maintain or change relations with them (Rothbart & Korostelina, 2006). It provides a sense of life and world, serves as a criterion for understanding actions and events, and regulates ingroup behaviors. With these criteria, individuals clarify group membership and relations with outgroups. "A collective axiology defines boundaries and relations among groups and establishes criteria for ingroup/outgroup membership. Through its collective axiology, a group traces its development from a sacred past, extracted from mythic episodes beyond the life of mortals, and seeks permanence" (Rothbart & Korostelina, 2006, p. 4). It is a set of constructions that are used to validate, vindicate, rationalize, or legitimize actions, decisions, and policies. Such constructions function as instruments for making sense of episodes of conflict and serve to solidify groups.

Two variables characterize the dynamics of collective axiology: the degree of collective generality and the degree of axiological balance.

1) *Collective generality.* The degree of collective generality "refers to the ways in which ingroup members categorize the Other, how they simplify, or not, their defining (essential) character" (Rothbart & Korostelina, 2006, p. 45). Collective generality includes four main characteristics:

   (i) homogeneity of perceptions and behaviors of outgroup members;
   (ii) long-term stability of their beliefs, attitudes, and actions;
   (iii) resistance to change;
   (iv) the scope or range of the outgroup category.

   A high level of collective generality is connected with viewing an outgroup as consistent and homogeneous, demonstrating fixed patterns of behaviors, committed to durable rigid beliefs and values, and widespread in the region or the whole world. A low degree of collective generality reflects the perception of the outgroup as differentiated, exhibiting a variety of behaviors, ready for transformation, and relatively limited in scope. An example of the high level of generality can be found in Greek history textbooks (see discussion below), which presents all Turks as homogeneous in their aggressive intentions, with a barbarian culture that dominates in society over centuries. An example of low-level generality is the transformation of history education in Germany that increases complexity in descriptions of the actions and motivations of ingroup and outgroups.

The degree of collective generality can change over time, especially in situations of growing intergroup tensions or violence. For example, the escalation of conflict can lead to the perception of an enemy not as a small local group but as an entire race, ethnic group, nationality, or culture. The image of an outgroup can become more rigid, firm, and homogeneous. During violent conflicts, people tend to deny the diversity and competing priorities within an outgroup and its multicultural and political structure, instead, perceiving it as a single "entity" with uniform beliefs and attitudes that support common policies toward other groups.

2) *Axiological balance.* "Axiological balance refers to a kind of parallelism of virtues and vices attributed to groups. When applied to stories about the Other, a balanced axiology embeds positive and negative characteristics in group identities" (Rothbart & Korostelina, 2006, p. 46). A balanced axiology leads to the recognition of decency and morality as well as immorality and cruelty among both the Other and the ingroup. A high degree of axiological balance reflects recognition of one's own moral faults and failings, while a low degree of axiological balance is connected with the perception of one's ingroup as morally pure and superior and of the outgroup as evil and vicious. This imbalance tends to promote a "tunnel consciousness" and a diminished capacity for independent thought.

"In its extreme form, a low axiological balance is correlated to exaggeration, inflation, and fabrication of outgroup vices and ingroup glories. The 'Them/Us' duality seems fixed in the timeless social order. With a fabricated sense of its collective virtues, the ingroup promotes a sense of moral supremacy over the outgroup. Such an unbalanced depiction of group differences provides a ground for a struggle against criminal elements of the world" (Rothbart & Korostelina, 2006, p. 47).

In education curricula, an unbalanced collective axiology and intolerance are developed through the presentation of the ingroup as peace-loving, moral, and victimized and the depiction of the outgroup as aggressive, vicious, and treacherous; the presentation of intergroup relations in terms of "ingroup victimization" – outgroup aggression.

Tolerance can be increased through an emphasis on a balanced collective axiology. In education curricula, a balanced collective axiology can be formed through the presentation of both positive and negative actions of the ingroup; critical analysis of political and social foundations and consequences of negative events; discussion of how aggressive actions of each side arrived from histories of intergroup relations; and reduction of negative and biased representation of outgroups.

Similarly, intolerance is connected with a high generality of collective axiology. In education curricula, high generality arrives from the absence of descriptions of differences in views and actions within both groups; emphasis on similarity of all members within the ingroup as well as within the outgroup; emphasis on permanence of the conflict between groups; and descriptions of the outgroup as always having aggressive intention and unable to change. Tolerance can be promoted through the formation of low generality of collective axiology that relies on the emphasis on differences within the ingroup and outgroup, diversity of opinions and view on conflict and intergroup relations, variety of extreme positions, and voices for tolerance; avoidance of presentation regarding the permanence of an outgroup's aggression through history; descriptions of positive change in relations; and descriptions of the outgroup as willing to reduce conflict.

4.10 Dehumanization

Dehumanization of outgroup members has two types: mechanistic and animalistic (Haslam, 2006). Mechanistic dehumanization rests on the denial of human attributes such as emotional responsiveness, interpersonal warmth, cognitive openness, and agency, which results in the perception of outgroups as cold, rigid, and machine-like. Animalistic dehumanization involves the denial of uniquely human attributes such as civility, refinement, and moral sensibility, leading to the perception of outgroup members as less human and more animal like. Animalistic dehumanization is also described as a process of "infrahumanization" that creates an underestimation of human emotions among outgroup members (Gaunt, Leyens & Demoulin, 2002; Gaunt, Leyens & Sindic, 2004; Leyens et al., 2000).

Both mechanistic and animalistic dehumanization are strongly connected with decreased tolerance toward the Other. In education curricula, mechanistic dehumanization is formed by descriptions of outgroups as cold-minded killers; depictions of outgroups as rigid and stubborn; and descriptions of outgroups as blind followers of leadership. Animalistic dehumanization can be promoted in education curricula through descriptions of outgroup members in animalistic terms; denial of morality in outgroup; and underestimation of human emotions among outgroup members.

4.11 Ideologization/Manipulation of Identity

Myths, as stories of origination, create the vision of the continuity of social community through a recounting of its past. They contribute to the salience of ingroup identity, delineate the meaning of ingroup membership, and establish the criteria for exclusion rooted in ingroup history and current position of groups within the society. Myths are contextualized within the political life of the community, providing a symbolic basis for social order, underpinning social interconnections, and legitimizing the existing social structure. It highlights and justifies the foundational norms and beliefs of a community, outlining and reshaping the connotations of social identity. Myths express the people's "reality postulates" about the world and concerns as "a moral universe of meaning" (Overing, 1997). Myths do not provide commemoration of mythical events — they reiterate them, making the protagonists of the myth present in contemporary life (Eliade, 1998).

Myths present significant features, norms, and beliefs of ingroups and outgroups, defining the boundaries between them and outlining rules of interaction. Myths contribute to the establishment of nations by determining their foundations, morality, and values. The historical validation of myth is not central to its meaning, rather, the core of myths is constituted by beliefs about criteria for goodness, legitimate participation, and exclusion/inclusion. Thus, myths are one of the crucial mechanisms of cultural reproduction and the "management of meaning" through the production and reproduction of significance in a particular context (Blumenberg, 1988; Bordieu, 1994; Horowitz, 1985; Smith, 2009, 2011). "A myth creates an intellectual and cognitive monopoly in that it seeks to establish the sole way of ordering the world and defining world-views. For the community to exist as a community, this monopoly is vital, and the individual members of that community must broadly accept the myth" (Schopflin, 1997; p.19). People sharing myths constitute a specific social community with a defined identity and social boundary, whereby all others are excluded.

There are three types of myths: foundational myths, legitimizing myths, and ideological myths. Foundational myths provide information about origins and the mission of a nation — they also define rights and obligations of different groups within a nation. Legitimizing myths are ideologies used by people to legitimate social hierarchies; ideological myths provide security, certainty, and moral authority. Among the 13 functions of the myth defined by Schopflin (1997), eight contribute to the development of social identity and five to the support of regime and legitimization of power. Thus, the first group of functions includes those of identity management that help to define and preserve common identity: 1) self-definition and self-attribution of the ingroup through the set of roles, functions, and purposes; 2) transference of identity and assimilation; 3) establishment of solidarity and illusion of the community through symbolic forms; 4) maintenance of collective memory; and 5) connection to culture. The second group includes functions related to intergroup relations and boundaries: 1) offering explanation for the fate of the community; 2) scapegoating; and 3) construction of the enemy. The third group includes functions of legitimization and support of power: 1) organizing and mobilizing public opinion; 2) simplification of complexity and standardization of knowledge; 3) transfer of political messages; 4) preservation of elites' power; and 5) assertion of legitimacy and strengthening of authority. The legitimizing function of myths is also analyzed through the approximation–creation of distant events closer to the group (Cap, 2007; Esch, 2010; Mazlish, 1981) or as a form of ideological control that maintains social systems and legitimizes power relations (Jost & Banaji, 1994; McFarland, 2005; Sibley & Duckitt, 2010; Sidanius & Pratto, 2001). The normative function of myths prescribes specific actions toward other nations (Korostelina, 2019).

Myths employ five mechanisms of justification: 1) impediment by the outgroup; 2) condemning imposition; 3) positive ingroup predispositions; 4) validation of rights; and 5) enlightening (Korostelina, 2013). They can be used in several types of myths or in a specific myth. The first justification mechanism, impediment by the outgroup, is the depiction of a fight between two groups in which the ingroup represents and supports positive values held by the nation. The desired values of the nation promoted by the ingroup vary from a mono-ethnic state based on nationalism to civic society and multiculturalism. The outgroup impedes ingroup activity through the development of conflict by establishing policies and promoting an ideology that is perceived to be wrong, and treating members of the ingroup unfairly through the use of oppression, and violence. Thus, the binary opposition between the "good' and "bad" groups is justified through the depiction of the right actions of the ingroup and the wrong actions of the outgroup. This mechanism posits ingroup exclusiveness in defining national identity and excludes the outgroup as an illegitimate agent of nation building and justifies the actions and dominance of the ingroup as representing the rightness in a nation. The impediment by outgroup mechanism can be more prominent in myths of foundation, suffering, and unjust treatment, and rebirth and renewal.

The second justification mechanism — condemning imposition — rationalizes the claim that the ingroup represents the interests of every group in the nation while the outgroup is imposing its own narrow ideology, ideas, policies, traditions, ethnic or regional culture, and language on everyone in the nation and wrongly claims to symbolize the nation. The myth explains why the culture or ideology of the outgroup is alien to the people and cannot be accepted by the nation. Thus, the binary opposition between "good" and "bad" groups is justified by the claim that the ingroup represents the whole nation, while the outgroup represents particular morally corrupt interests. This mechanism posits the ingroup as an essential core of the nation, while the outgroup is assigned to a narrow, corrupt subculture. It also justifies the power position of the ingroup in relation to all other groups. The condemning imposition mechanism can be more prominent in myths of ethnogenesis, territory, and Golden Age.

The third justification mechanism, positive ingroup predispositions, describes the ingroup as more able, capable, and competent than the outgroup. These abilities can include entrepreneurial ability and innovation, democratic values and cultures, tolerance, and the support of human rights. The myth describes them as stemming from a long history with greater development, which in turn becomes an essential core of the ingroup mentality. In comparison to the ingroup, the outgroup lacks these abilities because of its simplistic culture, regressive mentality, history, and geography of development. As a result, the outgroup is underdeveloped, conservative, and paternalistic but is trying to promote its ideas as those most suited for the nation. Thus, the ingroup is required to fight with a backward outgroup to prevent it from influencing national development. The opposition between "good" and "bad" groups is justified by the better abilities of one group to lead the nation. This mechanism posits the ingroup as progressive and virtuous, and, therefore, defending the nation from a backward outgroup and justifies the power of the ingroup as better able and suited to rule. The positive ingroup predispositions mechanism can be more prominent in myths of foundation and election.

The fourth justification mechanism, the validation of rights, describes the ingroup as having more rights to develop the nation according to their vision. These rights are based on a more advanced authentic culture, historic development on native land, birthright, and international acknowledgement. The outgroup has fewer entitlements because it is not native to the land due to its later arrival. The outgroup does not share ethnic roots with the ingroup and, as a result, is deemed to have a simplistic culture and cannot therefore be treated as an equal in the nation building process. In the extreme case — exclusion — the rights of the outgroup are completely denied and members are treated as alien and hostile and are excluded from the nation. The binary opposition between "good" and "bad" groups is justified by validating the exclusive rights of the ingroup and denouncing the rights of the outgroup. This mechanism posits the ingroup as legitimately deserving the power and the outgroup as alien to the nation. It justifies the power of the ingroup as coming from its history and rights to land. The validation of rights mechanism can be more prominent in myths of ethnogenesis and territory.

The fifth justification mechanism, enlightening, emphasizes the willingness of all people in a nation to pursue a particular goal, including civic society, liberalism, ethnic state, and multiculturalism but states that their limited abilities reduce their prospects to achieve their desired outcomes. Limitations stem from a persistent outdated mentality, absence of agency, and a dependency on populist leaders and government. The myth supports the claim of the ingroup as having a greater ability to identify the visions and aims shared by all and to enlighten them in their movement toward these goals. The binary opposition between "good" and "bad" groups is justified by positing the ingroup as the legitimate representatives of the nation, while people who do not share these visions are perceived as outsiders. This mechanism posits the ingroup as representing the shared vision of a positive future and the outgroup as not open-minded enough; it justifies the power of the ingroup as enlightened and progressive. The enlightening mechanism can be more prominent in myths of foundation.

Therefore, through the mechanisms of justification, mythic narratives serve to form and reestablish the specific meaning of national identity and legitimize the power of the ingroup, thus impacting the level of tolerance toward outgroups. In education curricula, impediments by an outgroup can be formed through descriptions of ingroups as having values of justice, equality, and liberty while the outgroup intends to destroy these values and descriptions of the outgroup as promoting destruction, injustice, inequality, and dictatorship. Condemning imposition can be promoted through the emphasis on the intention of the outgroup to totally destroy ingroup culture and values and the emphasis on assimilation policies of outgroups. Positive ingroup predispositions can be formed through descriptions of the ingroup as having more abilities and more competences than outgroups. The validation of rights and exclusion of an outgroup can be promoted through descriptions of ingroup as superior to outgroups or as one that should receive priority over others (coexistence is not mentioned/not considered as a viable option); depiction of ingroup culture (including religion) is superior to others or as one that should receive a priority. Finally, enlightenment can be formed in education curricula through descriptions of outgroups as wrong oriented; stress on the need to change the outgroup's view; and positioning that the ingroup has to teach/educate the outgroup.

4.12 The Conceptual Framework

Based on the previous discussion describing how the above concepts and theories can contribute to the formation of tolerance in education curricula, I propose the following framework for the assessment of tolerance development. Many existing research projects assess history education curricula through the binary concept of promoting peace/promoting violence. However, in conflicted and divided societies, history education faces the enormous task of addressing injustices, unbalanced power relations without provoking future violence. I propose a framework that progresses from 1) *incitement to violence and hatred* (intense dislike and hate of outgroup, justified willingness to fight with or harm outgroup members) to 2) *prejudice* (an unjustified or incorrect negative attitude toward members of outgroup based on the membership in ingroup), to 3) *tolerance* (as acceptance of the Other — fair, objective, and permissive attitude toward those whose opinions, beliefs, practices, racial or ethnic origins, etc., differ from one's own), and finally to 4) *mutual understanding* (including critical analysis of conflict, empathy, compassion, and willingness to cooperate), based on *14 indicators*, including salience and forms of identity, metacontrast, prototypes, favorable comparison, projection, social boundary, relative deprivation, collective axiology etc. (see Table 4.1).

This framework provides a more nuanced approach to forming tolerance through education as it takes into account gradual changes in curricula as well as the complexities of representations of traumas, power imbalances, and injustices of the past and current conflicts. It also provides an opportunity to have a multifaceted assessment of tolerance and reconciliation based on the 14 criteria. Some of them can be more developed toward mutual understanding, while others can be employed to promote hatred and intolerance. The framework can be used to map education curricula

Table 4.1 Identity-based conceptual framework for the assessment of tolerance in education curricula.

Concept	Mutual understanding (including critical analysis of conflict, empathy, compassion, and willingness to cooperate)	Tolerance (as acceptance of the Other — fair, objective, and permissive attitude toward those whose opinions, beliefs, practices, racial or ethnic origins, etc., differ from one's own)	Prejudice (an unjustified or incorrect negative attitude toward members of an outgroup based on the membership in an ingroup)	Incitement to violence and hatred (intense dislike and hate of outgroup, justified willingness to fight with or harm outgroup members)
Salience of identity	Promote multiple and cross-cutting identities	Promote multiple identities	Promote identities involved in conflict	Emphasize only conflict-related identity
Metacontrast	Very low level	Low level	High level	Very high level
Prototypes	Prevalence of peaceful prototypes	Both peaceful and aggressive	Prevalence of aggressive prototypes	Only aggressive prototypes
Favorable Comparison	Absent	Low level	High level	Prevalent
Projection	Absent	Some actions are justified through projection	Majority of actions are justified through projection	All actions are justified through projection
Global Attribution Error	Situational and dispositional attribution applies to both groups	Some prevalence of situational attribution to ingroup and dispositional attribution — to outgroup	Significant prevalence of situational attribution to ingroup and dispositional attribution — to outgroup	Situational attribution applies to only ingroup and dispositional attribution — only to outgroup

continued

Table 4.1 Continued

Forms of identity	Reflected	Cultural	Mix of cultural and mobilized	Mobilized
Social boundary	Positive narratives of boundary	Mostly positive narratives of boundary	Mostly negative narratives of boundary	Negative narratives of boundary
Relative deprivation	Analysis and critical assessment of inequalities	Emphasis on restoration of equality	Emphasis on relative deprivation	Emphasis on zero-sum competition
Chosen traumas and glories	Prevalence of glories, critical analysis of traumas	Equal presentation of traumas and glories	Emphasis on traumas, some glories support the idea of fighting	Presentation of only traumas, call for fighting
Threat	Critical analysis of threats	Some threats are emphasized with predominantly critical analysis	Emphasis on multiple threats	Significant emphasis on multiple threats
Collective axiology	Balanced with low level of generality	Mostly balanced with low level of generality	Mostly unbalanced with high level of generality	Unbalanced with high level of generality
Dehumanization	Absent	Low level of mechanistic dehumanization	High level of mechanistic dehumanization and low level of animalistic dehumanization	High level of mechanistic and animalistic dehumanization
Myths	Myths support foundations of coexistence, critical analysis of history	Some myths of enlightening and positive ingroup predispositions	Significant myths of enlightening, positive ingroup predispositions, and validation of rights	Prevalence of myths of impediment by outgroup, condemning imposition, and validation of rights

based on these 14 criteria and define concrete areas of improvement. It also can be used as a multidimensional tool for curricula revisions and the education of teachers to promote tolerance and mutual understanding among students.

References

Abrams, D. & Hogg, M. A. (1988). Comments on the motivational status of self-esteem in social identity and intergroup discrimination. *European Journal of Social Psychology, 18*, 317–334.

Adams, D. (2002). *The American peace movements.* Retrieved: http://www.culture-of-peace.info/apm/title-page.html

Bobo, L. D. (1999). Prejudice as group position: Microfoundations of a sociological approach to racism and race relations. *Journal of Social Issues, 55*, 445–472.

Bordieu, P. (1993). *The field of cultural production.* New York, NY: Columbia Press.

Boulding, E. (2000a). A New Chance for Human Peaceableness? *Peace and Conflict, 6*(3), 193–215.

Boulding, E. (2000b). *Cultures of Peace: The Hidden Side of Human History.* Syracuse, NY: Syracuse University Press.

Blumenberg, H. (1998). *Work on myth.* Cambridge, MA: MIT Press.

Blumer, H. (1958). Race prejudice as a sense of group position. *Pacific Sociological Review, 1*, 3–7.

Branscombe, N. R., Ellemers, N., Spears, R., & Doosje, B. (1999). The context and content of social identity threats" in N. Ellemers, R. Spears and B. Doosje (Eds.) *Social identity: Context, commitment, content.* Oxford, UK: Blackwell.

Branscombe, N. R., & Wann, D. L (1994). Collective self-esteem consequences of out-group derogation when a valued social identity is on trial. *European Journal of Social Psychology, 24*, 641–657.

Brewer, M. B. (2000). Superordinate goals versus superordinate identity as bases of intergroup cooperation," in D. Capozza and R. Brown (Eds.), *Social Identity Processes.* London: Sage.

Brewer, M. B. (2001). The many faces of social identity: Implications for political psychology. *Political Psychology, 22*, 115–125.

Brewer, M. B. (2007). The importance of being "we": Human nature and intergroup relations. *American Psychologist , 62*, 728–738.

Brighton, S. (2007). British Muslims, multiculturalism and UK foreign policy: "Integration" and "cohesion" in and beyond the state. *International Affairs, 83*, 1–17.

Brubaker, R. (1996). *Nationalism reframed: Nationhood and the national question in the new Europe.* Cambridge: Cambridge University Press.

Cadinu, M., & Reggiori, C. (2002). Discrimination of a low-status outgroup: The role of ingroup threat. *European Journal of Social Psychology, 32*, 501–515.

Cap, P. (2007). *Proximization in the discourse of politics: Legitimizing the 'war-on-terror'. Presented at the University of Colorado, Boulder, CO, USA.*

Cromwell, M. & Vogele, V.B. (2009). Nonviolent Action, Trust and Building a Culture of Peace. In J. De Rivera (Ed.), *Handbook on Building Cultures of Peace.* New York: Springer.

De Rivera, J. (2004). Assessing the basis for a culture of peace in contemporary societies. *Journal of Peace Research, 41*, 531–548.

Eliade, M. (1998). *Myth and reality.* Waveland Press Inc.

Esch, J. (2010). Legitimizing the 'war on terror': Political myth in official-level rhetoric. *Political Psychology, 31*(3), 357–391.

Esses, V. M., Dovidio, J. F., Jackson, L. M., & Armstrong, T. L. (2001). The immigration dilemma: The role of perceived group competition, ethnic prejudice, and national identity. *Journal of Social Issues, 57*, 389–412.

Esses, V. M., Haddock, G., & Zanna, M. P. (1993). Values, stereotypes, and emotions as determinants of intergroup attitudes" in D. M. Mackie, & D. L. Hamilton (Eds), *Affect, cognition, and stereotyping: Interactive processes of group perceptions.* San Diego, CA: Academic Press.

Fischer, P., Greitemeyer, T., & Kastenmuller, A. (2007). What do we think about Muslims? The validity of Westerners' implicit theories about the associations between Muslims' religiosity religious identity, aggression potential, and attitudes toward terrorism. *Group Processes & Intergroup Relations, 10*, 373–382.

Galtung, J. (1969). Violence, peace, and peace research. *Journal of Peace Research, 6*(3), 167–191.

Gaunt, R., Leyens, J. P., & Demoulin, S. (2002). Intergroup relations and the attribution of emotions: Control over memory for secondary emotions associated with ingroup or outgroup. *Journal of Experimental Social Psychology, 38*, 508–514.

Gaunt, R., Leyens, J. P., & Sindic, D. (2004). Motivated reasoning and the attribution of emotions to ingroup and outgroup. *International Review of Social Psychology,* 17, 5–20.

Gellner, E. (1994). "Nationalism and modernization," in J. Hutchinson and A. Smith (Eds.), *Nationalism.* Oxford: Oxford University Press.

Gurr, T. R. (1970). *Why men rebel.* London, UK: Routledge.

Gurr, T. & Harff, B. (1994). *Ethnic Conflict in World Politics* (2nd ed). Boulder, CO: Westview Press.

Hagendoorn, L., Linssen, H., Rotman, D., & Tumanov, S. (1996). *Russians as Minorities in Belarus, Ukraine, Moldova, Georgia and Kazakhstan* Proceedings from the International Political Science Association Conference, Boone, NC, USA.

Haslam, N. (2006). Dehumanization: An integrative review. *Personality and Social Psychology Review, 10*, 252–264.

Heider, F. (1958). *The psychology of interpersonal relations.* New York: John Wiley and Sons.

Herz, J. (1950). Idealist internationalism and the security dilemma. *World Politics, 2*, 157–180.

Hogg, M. A. (2000). Social identity and social comparison," in J. Suls and L. Wheeler (Eds.), *Handbook of social comparison: Theory and research.* Dordrecht, Netherlands: Kluwer Academic Publishers.

Hogg, M. A. (2007). "Uncertainty-identity theory," in M. P. Zanna (Ed.), *Advances in experimental social psychology.* San Diego, CA: Elsevier Academic Press.

Hogg, M. A., & Haines, S. C. (1996). Intergroup relations and group solidarity: Effects of group identification and social beliefs on depersonalized attraction. *Journal of Personality and Social Psychology, 70*, 295–309.

Hornsey, M. J. (2008). Social identity theory and self-categorization theory: A historical review. *Social and Personality Psychology Compass 2*, 204–222.

Horowitz, D. L. (1985). *Ethnic Groups in conflict.* Berkeley, CA: University of California Press.

Jervis, R. (1976). *Perception and misperception in world politics.* Princeton: Princeton University Press.

Jervis, R. (1978). Cooperation under the security dilemma. *World Politics, 40*, 167–214.

Jetten, J. & Spears, R. (2004). The divisive potential of differences and similarities: The role of intergroup distinctiveness in intergroup differentiation. *European Review of Social Psychology, 14*, 203–241.

Jetten, J., Spears, R., & Postmes, T. (2004). Intergroup distinctiveness and differentiation: A meta-analytic integration. *Journal of Personality and Social Psychology, 86*, 862–879.

Johnson, D., Terry, D. J., & Louis, W. R. (2005) Perceptions of the intergroup structure and anti-Asian prejudice among White Australians. *Group Processes & Intergroup Relations, 8*, 53–71.

Jones, E. E. & Harris, V. A. (1967). The attribution of attitudes. *Journal of Experimental Social Psychology, 3*, 1–24.

Jost, J. T. & Banaji, M. R. (1994). The role of stereotyping in system-justification and the production of false consciousness. *British Journal of Social Psychology, 33*(1), 1–27.

Iain, W. & Smith, H. H. (Eds.) (2001). *Relative deprivation: Specification, development and integration*. New York: Cambridge University Press.

Korostelina. K. V. (2019). The normative function of national historical narratives: South Korea perceptions of relations with Japan. *National Identities, 21(2)*, pp. 171–189.

Korostelina. K. V. (2013). *Constructing Narrative of Identity and Power: Self-imagination in a Young Ukrainian Nation*, New York, NY: Lexington

Korostelina. K. V. (2012). (Editor) *Forming a Culture of Peace: Reframing Narratives of Intergroup Relations, Equity, and Justice*. New York, NY: Palgrave Macmillan

Korostelina. K. V. (2007). *Social Identity and Conflict*. New York: Palgrave Macmillan.

Lake, D. & Rothchild, D. (1998). *The international spread of ethnic conflict: Fear, diffusion, and escalation*. Princeton: Princeton University Press.

Levine, R. A. & Campbell, D. T. (1972). *Ethnocentrism: Theories of Conflict, Ethnic Attitudes, and Group Behavior*. New York: John Wiley.

Leyens, J. P., Paladino, M. P., Rodriguez, R. T., Vaes, J., Demoulin, S. & Rodriguez, A. P. (2000). The emotional side of prejudice: The attribution of secondary emotions to ingroups and outgroups. *Personality and Social Psychology Review, 4*, 186–197

Louis, W. R., Duck, J. M., Terry, D. M., Schuller, R. A., & Lalonde, R. N. (2007). Why do citizens want to keep refugees out? Threats, fairness and hostile norms in the treatment of asylum seekers. *European Journal of Social Psychology, 37*, 53–73.

Mazlish, B. (1981). The next 'next assignment': Leader and led, individual and group. *The Psychohistory Review, 9*(3), 214–237.

McFarland, S. G. (2005). On the eve of war: Authoritarianism, social dominance, and American students' attittudes toward attacking Iraq. *Personality and Social Psychology Bulletin, 31*(3), 360–367.

Michael J. A., Wohl, M. J. A., Nyla, R., Branscombe, N. & McVicar, D.N. (2001). "One day we might be no more": Collective angst and protective action from potential distinctiveness loss. *European Journal of Social Psychology, 41*, 289–300.

Musgrove, L., & McGarty, C. (2008). Opinion-based group membership as a predictor of collective emotional responses and support for pro- and anti-war action. *Social Psychology, 39*, 37–47.

Neisser, U. (1967). *Cognitive psychology.* New York: Appleton-Century-Crofts.

Nicholson, H. (2014). Social identity processes in the development of maximally counterintuitive theological concepts: Consubstantiality and no-self. *Journal of the American Academy of Religion, 82*(3), 736–770.

Noor, M., Brown, R., Gonzalez, R., Manzi, J., & Lewis, C. A. (2008). On positive psychological outcomes: What helps groups with a history of conflict to forgive and reconcile with each other. *Personality and Social Psychology Bulletin, 34* , 819–832.

Quillian, L. (1995). Prejudice as a response to perceived group threat: Population composition and anti-immigrant and racial prejudice in Europe. *American Sociological Review, 60*, 586–611.

Overing, J. (1997). The role of a myth: An anthropological perspective" in G. A. Hosking and G. Scöpflin (Eds.), *Myths & nationhood.* New York, NY: Taylor & Francis.

Pettigrew, T. F. (2015). Samuel Stouffer and relative deprivation. *Social Psychology Quarterly, 78*(1), 7–24.

Pettigrew, T. F. & Tropp, L. (2011). *When groups meet: The dynamics of intergroup contact.* New York: Psychology Press.

Posen, B. (1993). The security dilemma and ethnic conflict. *Survival, 35*, 27–47.

Reynolds, K. J., Turner, J. C., & Haslam, S. A. (2000). When are we better than them and they worse than us? A closer look at social discrimination in positive and negative domains, *Journal of Personality and Social Psychology, 78*, 64–80.

Richards, H. & Swanger, J. (2009). Culture Change: A Practical Method with a Theoretical Basis. In J. De Rivera (Ed.) *Handbook on Building Cultures of Peace.* New York: Springer.

Ross, L. (1977). The intuitive psychologist and his shortcomings: Distortions in the attribution process" in L. Berkowitz (Ed.), *Advances in Experimental Social Psychology*. New York: Academic Press.

Rothbart, D. & Korostelina, K. V. (2006). *Identity, morality and threat*. Lexington, MA: Lexington.

Sears, D. O., & Henry, P. J. (2003). The origins of symbolic racism. *Journal of Personality and Social Psychology, 85*, 259–275.

Schopflin, G. (1997). The functions of myth and taxonomy of myths" in G. A. Hosking and G. Scöpflin (Eds.), *Myths & nationhood*. New York, NY: Taylor & Francis.

Sherif, M. (1966). *Group Conflict and Cooperation: Their Social Psychology*. London: Routledge and Kegan Paul.

Sherif, M. & Sherif, C. (1953). *Group in Harmony and Tension*. New York: Harper.

Shin, J. Y. (2018). Relative deprivation, satisfying rationality, and support for redistribution. Social Indicators Research, *140*(1), 35–56.

Sibley, C. G. & Duckitt, J. (2010). The ideological legitimation of the status quo: Longitudinal tests of a social dominance model. *Political Psychology, 31*(1), 109–137.

Sidanius, J. & Pratto, F. (2001). *Social dominance; An intergroup theory of social hierarchy and oppression*. Cambridge, UK: Cambridge University Press.

Smith, A. D. (2009). *Ethno-Symbolism and Nationalism: A Cultural Approach*. London; New York: Routledge.

Smith, A. D. (2011). National identity and vernacular mobilisation in Europe. *Nations and Nationalism 17*(2), 223–256.

Smith, H., Pettigrew, T. F., Pippin, G. & Bialosiewicz, S. (2012). Relative deprivation: A theoretical and meta-analytic critique. *Personality and Social Psychology Review, 16*, 203–32.

Snyder, J. & Jervis, R. (1999). Civil war and the security dilemma" in B. F. Walter and J. Snyder (Eds.), *Civil wars, insecurity, and intervention*. New York: Columbia University Press.

Solomon, G. & Cairns, E. (2009). *Handbook on Peace Education*. New York: Psychology Press.

Spear, J. (1996). Arms Limitations, Confidence-Building Measures, and Internal Conflict" in M. E. Brown (Ed.), *The international dimensions of internal conflict*. Cambridge, MA: MIT Press.

Staub, E., Pearlman, L. A., & Hagengimana, A. (2005). Healing, reconciliation, forgiving and the prevention of violence after genocide or mass

Killing: An intervention and its experi- mental evaluation in Rwanda. *Journal of Social & Clinical Psychology, 24*, 297–334.

Stephan, W. G., Boniecki, K. A., Ybarra, O., Bettencourt, A., Ervin, K. S., & Jackson, L. A., (2002). The role of threats in the racial attitudes of blacks and whites. *Personality and Social Psychology Bulletin, 28*, 1242–1254.

Strabac, Z., & Listhaug, A. (2008). Anti-Muslim prejudice in Europe: A multilevel analysis of survey data from 30 countries. *Social Science Research, 37*, 268–286.

Strelan, P. & Lawani, A. (2010). Muslim and Westerner Responses to Terrorism: The Influence of Group Identity on Attitudes Toward Forgiveness and Reconciliation. *Peace and Conflict, 16*, 59–79.

Tajfel, H. & Turner, J. C. (1979). "An Integrative Theory of Intergroup Conflict," in W. G. Austin and S. Worchel (Eds.), *The Social Psychology of Intergroup Relations*. Monterey: Brooks/Cole.

Taylor, D. M. & Moghaddam, F. M. (1994). *Theories of Intergroup Relations: International Social Psychological Perspectives (2nd ed)*. New York: Praeger.

Turner, J. C. (1975). Social comparison and social identity: Some prospects for intergroup behavior. *European Journal of Social Psychology, 5*, 5–34.

Turner, J. C. (2000). "Social identity," in A. E. Kazdin (Ed.), *Encyclopedia of psychology, volume 7*. Washington, DC: American Psychological Association.

UNESCO (1995). *UNESCO and a Culture of Peace: Promoting a Global Movement*. New York: United Nations Educational, Scientific, and Cultural Organization.

Van Evera, S.. (1999). *Causes of war: Power and the roots of conflict*. Ithaca: Cornell University Press.

Volkan, V. (1998). *Bloodlines: From ethnic pride to ethnic terrorism*. New York, NY: Basic Books.

Warburg, B. (2010). Germany's national identity, collective memory, and role abroad" in E. Langenbacher and Y. Shain (Eds.), *Power and the past: Collective memory and international relations*. Washington, DC: Georgetown University Press.

Wheeler, N. J. & Booth, K. (1992). The Security Dilemma" in J. Baylis and N. J. Rengger (Eds.) *Dilemmas of world politics: International issues in a changing world*. Oxford: Clarendon Press.

Wohl, M. J. A., & Branscombe, N. R. (2005). Forgiveness and collective guilt assignment to historical perpetrator groups depend on level of social category inclusiveness. *Journal of Personality and Social Psychology, 88,* 288–303.

Wohl, M. J. A., & Branscombe, N. R. (2009). Group threat, collective angst and ingroup forgiveness for the war in Iraq. *Political Psychology, 30,* 193–217.

Wohl, M. J. A., Branscombe, N. R., & Klar, Y. (2006). Collective guilt: Emotional reactions when one's group has done wrong or been wronged. *European Review of Social Psychology, 17,* 1–36.

5

The Historical-Legal Development of Religious Tolerance and Harmony in Albania

Eda Çela

University of Elbasan

There are three identifying cultural values of Albanian society that have survived and coexisted for years. They are hospitality, besa[1] (the given word), and religious coexistence. These values have survived the challenge of globalization, but, nowadays, they are at constant risk of being misused.

The case of Albania is also special because in a homogeneous ethnic composition, like few countries in the Balkans, religions in Albania have been diverse. In this way, Albania appears as a nation and as a multi-religious society. This has created what is considered the Albanian religious identity.

But this is not a sporadic identity.

The year 1912 is considered to be the year of hope and great changes for Albania and Albanians. On 28 November 1912, Albania declared independence. Until this time, it had been under the sovereignty of Turkey. A regulatory legal framework was also being adopted at this time. In the north of Albania, the Kanun of Leke Dukagjini was the only legal act regulating all social, economic, and coexistence rules in the area. The activity of Kanun was very limited. In the other part of Albanian territory, there started to emerge the legal framework adopted by the national assembly of Vlora. At the same time, it was held during the London Conference which redefined the borders of the Balkans and, consequently, the borders of Albania. Among other things, it was urged to be aware of the adoption of the legal framework, with particular attention of preventing the spread of belonging and Muslim

---

[1] A promise made by Albanians which is destined to fulfill.

communities in the country.[2] At that time, it was very important to maintain public order, regardless of how relations between religious communities would be regulated. The Organic Statute of Albania was approved on 10 April 1914 in Vlora.[3] In order to preserve the religious balance in the country, the statute sanctioned that Albania has no official[4] (fe shtetërore) religion. The freedom and public exercise of all cults are guaranteed, even nowadays.

There are specific provisions of this law which regulate relations with religious communities. All existing Albanian religious communities are recognized (by the state). This principle also applies to the various Muslim sects. On the other hand, there must be no obstacles regarding the hierarchical organization of various communities, nor to the relationships these communities might have on religious dogma with their higher spiritual leaders. Religious communities have always protected and maintained their properties, and their buildings have always been excluded from tax liabilities.

The 1920 Statute of Lushnja,[5] supplemented by the expanded Statute of Lushnje, 1922, which was the basic law, provided general rules regarding the functioning of the state. Among other things, in a separate chapter which summarizes the various provisions, the Statute declared that there is no official religion in Albania.[6] This provision is of particular historical legal relevance as it is an accepting and not prohibitive provision. This means that the state has no official religion, but not the citizens.

The legislator, in the continuation of the provision, also developed acceptance, coexistence, and other religions. All religions are honored, accepting their diversity and coexistence. This prediction also creates a division between religion and faith. In this way, all religions, meaning all organized systems of beliefs, ceremonies and rules are accepted to worship God and regulate human relations. According to that, the provision also regulates the freedom to practice certain religions and faiths. The legislator

---

[2] Giovanni Cimbalo, Pluralismo confessionale e comunità religiose in Albania, Bononia University Press, ISBN 978-88-7395-762-1, pg 22.

[3] Albania is constituted as a constitutional, sovereign, and hereditary principality under the guarantee of the six great powers, headed by Prince Vid.

[4] Legally accepted by the government.

[5] This act is considered by many scholars as the first Constitution of the Republic of Albania, in terms of its form and content.

[6] Article 93 of Elected Statute of Lushnja, 8-12-1922.

states that freedom to exercise their religion and beliefs are guaranteed to everyone.

Religious and belief restraint extends its effects on the organization and functioning of the state. Religion cannot become a legal barrier, nor can it be used for political purposes. It is the law itself that has foreseen the limits of activity, limiting particulars that violate state activity. This legal provision was not only advanced for the time, but, on the other hand, it also laid down the legal basis for the guarantee of religious harmony and coexistence in Albania.

But what is more important is the recognition of the full autonomy of the religious communities, proving that "the heads of every religion and every confession are chosen on the basis of their statutes, and recognized by a decree by part of the government" which confirms the jurisdictional approach of the law. "Every religious association and institute like: a mosque, a Teqe, a Church, a Monastery etc., is recognized as a legal person and is represented on the basis of one's own religious rules," with an explicit postponement to the internal ordering of cults. There could be of no greater guarantee than this to testify to the effective recognition of the religious freedom of associations and of individuals, especially since different cult organizations have the right to have, acquire, dispose of, and administer movable and immovable assets.[7]

The approval of the Statutes of the various religious communities, imposed by the constitutional provisions referred to above and by international events, leads in 1923 upon the approval of the law on the legal status of religious communities with which confers civil juridical personality on the various cults.[8] Thus, in 1923, Albania — the only state in Europe — adopted legislation on the attribution of legal personality to religious communities which represents the most advanced point of a long elaboration of European law, proving to be at the center of the debate on what the peculiar characteristics, the formants must be, necessary for the construction of a modern state.[9]

As the only state in Europe — legislation on the matter of attribution of juridical personality to the religious communities, representing the most

---

[7] Ibid.

[8] Legal Statute of Religious Communities, http://licodu.cois.it/415/view.

[9] Giovanni Cimbalo, Pluralismo confessionale e comunità religiose in Albania, Bononia University Press, ISBN 978-88-7395-762-1, pg. 37.

advanced point of a long elaboration of European law — Albania performed the characteristics for the construction of a modern state.[10]

In the same line, when the form of state changed in Albania, from a parliamentary Republican state,[11] in a democratic, parliamentary, and hereditary kingdom state. On 1 December 1928 was approved the Basic Statute of the Albanian Kingdom, headed by King Zog. This act was similar in form and content to all the constitutions of that time. In the preamble of this act are invoked the values of national unity, the assurance of peaceful development of the homeland, and the common good while respecting historical traditions. This preamble glorifies the values of the Albanian people who, for religious reasons, appear to be united.

In the same form as the Extended Statute of Lushnja, the Statute of the Albanian Kingdom institutionalized that the Albanian State had no official religion. All religions and beliefs are honored and the freedom to practice each is guaranteed. Religion cannot form legal barriers. Religions and beliefs can never be used for political purposes.[12]

It is an interesting fact that this law provided the incompatibility of the function of the deputy with people with active religious service. Over the years, the most educated and intellectual parts of the Albanians have had a genuine religious orientation. For this reason, in delegating bodies and assemblies, delegates belonged to different religious communities, in the capacity of representing the best interests of the population. They had a very high reputation, and their word was often decisive in decision making. Such was the national assembly held on 28 November 1912 in Vlora, consisting of 83 delegates of the Muslim and Christian faiths, who signed the Albanian Declaration of Independence. In 1924,[13] Albania's Prime Minister was Fan Noli, who previously served as a priest. He continued to be a priest even after the fall of the government of 1924.

---

[10]Ministero per i beni e le attivita culturali, Direzione generale per gli archivi, *L'unione fra l'Albania e l'Italia*, a cura di S. Trani, Roma, Pubblicazioni degli Archividi Stato, Strumenti CLXXIII, 2007, 72–82,

[11]The proclamation Republic of Albania, for the first time in history, took place on 21 January 1925, on the 5th anniversary of the beginning of the Lushnja Congress.

[12]Article 5 of the Basic Statute of the Kingdom of Albania, 01 December 1928.

[13]16 June 1924 to 24 December 1924 has been in Albania what is known in history as the Fan Noli Government. He was the first Prime Minister to come from the ranks of Catholic clergy in Albania. In 1908, he was appointed (ORDAINED) priest and founded the Albanian Church.

Meanwhile, since 1928, this situation has changed. First, part of the intellectual and well-educated elite in Albania did not belong only to those who were in active religious service. Second, the mentality of the people who demanded that their representatives not only belong to religious beliefs had changed, but there was a general inclusion of the representatives' categories.

The Statute of the Kingdom went further, stating that the Member of Parliament could not attend Parliamentary meetings in religious clothing. In this way, the manifestation of a certain religious belief was left out of the doors of the legislative body.

I think it is important to mention in this paper the formula of the oath in case of taking the mandate of a deputy or a king.

Before committing to office, MPs swear this way: "I swear in the name of God, that as an Albanian MP I will be faithful to the Statute, I will work honorably and conscientiously for the good of the Fatherland."[14]

Also the formula of the King's oath in Parliament was conceived in this way: "I (.......), the King of the Albanians, as ascends the throne of the Albanian Kingdom and takes over the royal powers, I swear before the Almighty God, that I will preserve national unity, state independence and territorial integrity; I will also abide by the Statute and act in accordance with it and the laws in force, always taking into account the good of the people. God help me."

Although a laic state, before taking up representative duties in both cases, there is a direct reference to "God." This shows that not only for the sake of a legal provision but also because of a self-accepting culture, Albanians have accepted the coexistence and harmony of their faith and religion.

Religious coexistence and harmony in Albania at that time continued to be complemented by a rich legal framework. For the first time in Albania, there was an adoption of special legislation on religious communities. The law on the administration of religious communities of 01.02.1930, among others, quoted that: "Religious communities in Albania were moral legal entities who enjoyed all rights except to participate in political activities directly or indirectly. Religious communities enjoyed the rights conferred by law only when the Statute presented by them was approved by the Council of Ministers and decreed by the King."

Eqerem Bey Vlora, a scholar, among others referring to the period of King Zog, has quoted: "Zog created not only a state but also a nation."

---

[14] Article 30 of The Basic Statute of the Kingdom of Albania, 01 December 1928.

This philosophy was also served by the relationships he instituted with the religious communities in Albania.

5.1 Communist Albania and Forced Atheism

The elements of historical legal treatment of religious tolerance in Albania relate to the most important moments selected to support this thesis.

In October 1944 in Berat was held the second meeting of the Anti-Fascist National Liberation Council, which decided to transform and mandate Albania into temporary democratic government. A Declaration of Civil Rights was adopted at this time, which guaranteed the exercise of freedoms, universal rights, and the right to religious belief and conscience, as equal rights are accorded to all religious beliefs. For the moment, there were no significant changes in religious life; indeed, it reaffirmed the equal treatment between the confessions: the politics of national unity required it. Resistance to religion would emerge later.

In 1945, based on the law on the Constituent Assembly and the electoral lawm sanctioned the right to vote without distinction of religion and political ideas, of women and of all citizens over the age of 18. The elections of 2 December 1945 saw the victory with 90% of the Democratic Front dominated by the communists; the government presided over by Enver Hoxha took office on 24 March 1946.

Meanwhile, the Constitution was promulgated on 14 March 1946 which reaffirmed the state to be laic and teaching; ownership and private initiative were guaranteed, even if it was established that private property can be limited and expropriated when the general interest requires it and always according to the law.

An entire chapter of the Constitution, the 13th, was dedicated to the rights and obligations of citizens, who were guaranteed freedom and equality in religious matters and the separation between state and religions was sanctioned, even if the state could economically support the confessions. The law on political parties,[15] in Article 7, prohibited those with a religious background and the Government urged the religious communities to re-establish the ecclesiastical hierarchies that would be left free to carry out their activities according to the legislation and statutes approved in the period of King Zog. The approval of the law marks the end of a first phase of relations in the

---

[15]Decree no. 241 dated 06 December 1946 "On the establishment of political parties", Official Gazette no. 115, dated 23 December 1946.

name of the reconstruction of national unity, after which a phase opens in the relations between the political authorities and the various religious communities which will last until 1967, during which the government subordinates the very existence of religious communities to the building of the socialist state and will order its suppression.

According to Prof. Hysi,[16] from a 1945 survey, 68%–70% of Albanians were Muslims, 17% Orthodox, and more than 10% Catholics. In addition to the Sunnis, who were the majority, five other brotherhoods were included within the Muslim community: haleveti, rufai, kadri, tixhan, and sadi. The Bektashi maintained their own organization.

In the first years of communist governance, the modalities of application of the statutory norms constituted an internal matter of religious communities. The war and the struggle for the liberation of the country had involved many members of the clergy or militant members of the religious communities and these wished to bring elements of democratization and participation within the communities themselves; therefore, they fueled the confrontation within the respective religious organizations.

From the speech of the Communist dictator, held on 6 February 1967, began the most tragic stage of an entire Albanian nation, Calvary, to the Crucifixion and, with God's help, to the Resurrection. As history later showed, atheistic communism educated a whole generation without conscience, that is, inhuman ideals, with disrespect for the person, with disrespect for the freedom of the other, instilled a spirit that was neither spirit nor Albanian national tradition, nor is the tradition of European humanism[17]. Enver Hoxha delivered a speech entitled "Further Revolutionizing Party and Power," a speech in which he began the fight against "religious ideology" and "customs," which later led to the proclamation of Albania as the first and the only atheist country in world history.[18]

On 28 December 1976,[19] it was approved by the Constitution of the Popular Socialist Republic of Albania. In Article 37, it was stipulated that:

---

[16]ShyqyriHysi, Myslimanizmi në Shqipëri 1945–1950, Tirane, Mesonjtorja e pare, 2000, 7.

[17]Martirët e Lum të Kishës katolike në Shqipëri Vatican news, 06 February 2019

[18]With a secret order of dictator Enver Hoxha in 1967, figures and details are given on the destruction of cult sites in Albania and the open fight against religious beliefs. This number totals 2169, of which 740 mosques, 608 Orthodox churches and monasteries, 157 Catholics, 530 teqes, mausoleums etc.

[19]Law nr. 5506 dated 28 December 1976 The Constitution of the Popular Socialist Republic of Albania.

"The state does not recognize or support any religion and develops atheistic propaganda to instill in people the materialistic scientific understanding."

Meanwhile, Article 55 of the Criminal Code of 1977 in Albania stated:

"Fascist, anti-democratic, religious, belligerent, anti-socialist agitation and propaganda, as well as the preparation, dissemination or preservation for the dissemination of literature with such content to weaken or undermine the state of the proletariat dictatorship, is punishable by deprivation of liberty from three to ten. After these offenses, when committed in times of war or have caused particularly grave consequences, are punished: with deprivation of liberty not less than ten years or by death."

Albania became the first atheist country in the world. This was the decision taken by the communist government. After over 50 years of faith in God, religions of any form of manifestation of faith were then completely banned and even barbarously condemned.

The consequences were not small in terms of ideology. It is a twin generation who grew up totally distanced from religious beliefs or practicing it. This made Albanians indifferent to religion-related issues but, at the same time, fueled by a sense of hatred and fight against preachers who practiced religion in secret. The fact that the cult sites collapsed and were destroyed showed a propaganda of hatred and the fight against religion.

According to Sokol Paja,[20] media in totalitarian Albania totally serves this totalitarian country. They used to transmit information that is compatible with the political and ideological principles of this state. The power itself, media of that time and community, functions as a unit. The atheist propaganda of communist system had a special importance regarding the awareness due to the provisions taken in order to fight the proletariat, the proletarian triumph, and settling the dictatorship in Albania, and, of course, media has a specific role.

5.2 The Reborn of Religious Tradition, Tolerance and Harmony in Albania

The collapse of communist systems in Eastern Europe brought about a new system of new organizational spirit in Albania. On 29 April 1991, Law No. 7491 "On the Basic Constitutional Provisions" was adopted, which played

---

[20]Ph.D. candidate, Universiteti Tiranes, Departamenti I Gazetarisë dhe Komunikimit, Book of Abstracts, First International Conference on Communication and MediaStudies, 29–30 May 2015, Kolegjia AAB, Pristina, Kosovo, pg. 44.

the role of a provisional constitution. Meanwhile, state-owned activities and undertakings for Albania's accession to the Council of Europe had begun, in consultation with constitutional initiatives and with the Venice Commission.

The new Constitution certainly meets the criteria set by the Council of Europe for the countries that intend to be part of this organization. The legal nature and the freedom of religious communities, is the protection of individual and collective religious freedom, because in fact it is in a line of continuity with the legal experiences that have characterized the Albanian state since its independence.

In fact, the laic nature of the state, already present in the Statute of 1914, is reconfirmed as the regime of recognition of the juridical personality to which religious communities are subjected. Thus, the Constitution moves in the wake of the legal tradition of the country that, since 1923, has adopted pluralism of cults as one of its distinctive characters, a choice that makes Albania a unique case among the countries where Islam is in the majority.

This allows us to argue that Albanian is a European Islam quite different from that practiced in other areas of the world. These particular parameters of legal and institutional civilization, as well as attention to religious pluralism and secularism, confer on the Albanian system a capacity for social stabilization which has avoided Albania's ethnic-religious wars that have characterized other countries.

Conversely, the absence of interreligious conflicts, together with national cohesion, has played a fundamental role in keeping out the populations allocated on the territory of the Albanian state from the wars of religion, so frequent in the western Balkans.

However, after the fall of the regime and the absence of legal instruments in the system to manage and regulate the activities of the religious communities, they try to present themselves in another form or to use institutions set up for other types of activities that could somehow cover the void existing legislation and allowed them to operate in the territory of the state.

Article 7 of the law for the constitutional provisions of 1991, stipulated that:

The Republic of Albania is a secular state.

The state respects the freedom of religion and creates the conditions for its exercise.

Following the adoption of the law, a number of religious communities in Albania sought to organize their registration as organized formations. Some courts agreed to do so and others found the legal framework insufficient.

However, in the new institutional framework, the existence of religious communities and the individual protection of religious freedom is again guaranteed at the level of constitutional principles, and this is soon reflected in the new civil code, published in 1994,[21] whose article 39 establishes that: associations are social organizations that pursue a political, scientific, cultural, sporting, religious, charitable, or any other non-economic purpose, thus making religious communities fall among those social formations whose activities are regulated by common law legislation.

These guarantees provided with regard to the rights of freedom of association are accompanied by the restoration of the activities of religious communities through the development of procedures for the restitution to them of expropriated properties from 1945 and definitively confiscated in 1967.

In 1998, the constitution of the Republic of Albania was adopted, which also passed the referendum filter.[22] Unlike any previous constitution or act with the power of the constitution, the 1998 constitution, in its preamble, states: "We, the people of Albania, are proud of our history,... with a spirit of religious coexistence and tolerance, ... with a deep conviction that justice, peace, harmony and cooperation among nations are among the highest values of humanity......."

In fact, the constitution, as the highest legal act in the Albanian legal system, has made its content subject to a number of articles that support and teach the harmony of religious tolerance in Albania.

Article 3[23] lists the basic principles on the basis of which the Albanian state was founded, among other things religious affiliation and coexistence. It also states that it is the duty of the state to respect and protect them.

---

[21] Civil Code of the Republic of Albania, approved by law. 7850 dated 29 July 1994

[22] Adopted by law no. 8417, dated 21 December 1998 of the People's Assembly. Approved by referendum on 22 November 1998 Declared by Decree no. 2260, dated 28 November 1998 of the President of the Republic, Rexhep Mejdani Amended by Law no. 9675, dated 13 January 2007 Amended by Law no. 9904, dated 21 April 2008 Amended by Law no. 88/2012, dated 18 September 2012 Amended by Law no. 137/2015, dated 17 December 2015 Amended by Law no. 76/2016, dated 22 July 2016.

[23] Article 3 of the Constitution of the Republic of Albania: Independence of the state and the entirety of its territory, human dignity, its rights and freedoms, social justice, constitutional order, pluralism, national identity and national heritage, religious coexistence and coexistence, and the understanding of Albanians with minorities are the basis of this state, which has a duty to respect and protect them.

Religious communities are considered part of the social formations on which the Albanian state is based. Article 9 of the Constitution provides for an exception to the first rule: the foundation of political parties and other organizations having totalitarian programs, activities and methods, which instigate or defend religious, racial and territorial hatred, which use the violence to govern or make politics in the country, thus reproducing more broadly a rule that has historically been part of the country's constitutional provisions.

As a consequence, the Constitution has dedicated a full article to religious matters.

Article 10:

1) There is no official religion in the Republic of Albania.
2) The state is neutral in matters of faith and conscience and guarantees the freedom of expression in public life.
3) The state recognizes the equality of religious communities.
4) The state and religious communities respect each other's independence and cooperate for the benefit of each and every one.
5) Relations between the state and religious communities shall be governed by the agreements reached between their representatives and the Council of Ministers. These agreements are ratified by the Assembly.
6) Religious communities are legal entities. They are independent toward the administration of their property according to their principles, rules and laws, insofar as the interests of third parties are not affected.

In this way, the traditional legislation on religious communities is restored which, as we have seen, has distant and thoughtful origins in Albanian law. With this provision, the atheist parenthesis introduced with the suppression of religious communities by the 1967 decree closes and resumes the legal tradition relating to the activity of religious communities with civil legal personality.

Article 18 of the Constitution establishes a fundamental principle — that of equality. All are equal before the law, without distinction of sex, race, religion, ethnicity, language, political, religious and philosophical opinion, economic, social situation, right to education, and parental belonging. Nobody can be discriminated against for the reasons just mentioned, if not for justified reason.

The State guarantees, pursuant to Article 20, the rights of ethnic minorities who enjoy equal rights and freedoms before the law, can freely express their ethnic, cultural belonging, receive education in their mother tongue, and can meet in organizations and groups to defend their interests and their identity.

Article 24 attributes to the State the role of guarantor of the right to declare and practice publicly the religion of belonging. The Constitution has the merit of balancing the principle of impartiality of the state in religious matters and that of state protection of the right of religious communities to participate in the public life of the country.

On an individual level, Article 24 guarantees the freedom of conscience and religion and the right to choose, change, and manifest one's religion and beliefs, privately or publicly, through worship, education, and the celebration of religious functions. When it states in point 24.3 that "no one can be forced or deprived of the right to participate in a religious community or its practices" the Constitution takes up the same provision contained in Constitutions such as the Czech or Lithuanian, but common to many Eastern countries.[24]

Religious communities in Albania, as legal entities, have independence in the administration of their properties according to their principles, rules, and canons, relations between the state and religious communities are regulated on the basis of agreements. They can also recourse the Constitutional Court regarding the issues connected to their interests.[25]

Albanian legislation is rich in laws and bylaws that regulate in detail the religious conventions but are also an expression of tolerance, acceptance, and harmony in the country. By law, the Office of the Commissioner for Protection from Discrimination receives and processes discrimination complaints, including those related to religious practices. The law specifies that the State Committee on Cults, under the jurisdiction of the Office of the Prime Minister, regulates relations between government and religious groups, protects freedom of religion, and promotes interfaith cooperation and understanding. The law also instructs the committee to keep records and statistics on foreign religious groups seeking help, and to support foreign religious group workers in obtaining residence permits.

The law allows religious communities to run educational institutions as well as to build and administer religious cemeteries on the lands that those

---

[24]Giovanni Cimbalo, Pluralismo confessionale e comunità religiose in Albania, Bononia University Press, ISBN 978-88-7395-762-1, pg. 129.

[25]Even though they have this right, they have not yet brink a case to the Constitution Court.

communities own. In 2009,[26] a special law was adopted on state budget financing religious communities.

5.3 Religious Harmony in Albania, Between Myths and Truth

Religious harmony in Albania is considered an undeniable value for every Albanian. This value has its roots in the Albanian society as such, rather than a value created by its practical application. Albanians accept religious harmony in terms of a belief, cultural, and social policy in the country, despite the historical, legal, and political changes in the country.

Albanians lead a secular, legal-oriented life but, on the other hand, have a very high level of tolerance for the organization and functioning of the underlying communities in the country, accepting new communities without any differentiation or conditionality compared to other communities.

What is very important to note is the fact that Albanians acknowledge religious diversity but do not accept that one religion will prevail over another. This is a very interesting element of harmony in Albania, as avoiding the hierarchy of one religion over another, all acceptance is developed and promoted as a distinctive feature.

In the cultural aspect, religion occupies a very important place in the manifestation of religious beliefs through the holidays. In Albania, every holiday of every religious faith is celebrated. Albanians also celebrate the holidays of their individual religious beliefs as well as those of others. Currently, in Albania, the following are celebrated: Nevruz Day, Catholic Easter Sunday, Orthodox Easter Sunday, Eid al-Adha, Eid al-Adha Day, Mother Teresa's Day of Christmas, Christmas, and Ramadan Month. These celebrations are peaceful and in mutual respect for the rituals of everyone.

Given the historical legal development of religious tolerance in Albania, maintaining social cohesion and religious tolerance and harmony, even in turbulent times, has been a remarkable achievement.[27]

So far, the Albanian state has signed five agreements with the communities: Catholics, Muslims, Orthodox, Bektashi, and Evangelicals. The

---

[26]Law No.10 140, dated 15 May 2009 on the financing from the state budget of the religious communities, which have signed agreements with the council of ministers.

[27]Study on Religious Tolerance in Albania, UNDP, Programi I Kombeve te Bashkuara per Zhvillim, 2018, https://www.undp.org/content/dam/albania/docs/religious%20tolerance%20albania.pdf

institutionalization of these reports has also contributed to religious tolerance and harmony in Albania.

The Religious Tolerance Report in Albania[28] has concluded that the foundations of religious tolerance in Albania are deeply rooted in social traditions and culture and do not come from the consciousness, knowledge, or practice of religious rites. As they recognize the role of religious leaders in fostering religious tolerance nowadays and throughout history, Albanians place greater importance on the legal and practical separation between state and religion; respect for human rights and fundamental freedoms; secularism as a feature of society; and national sentiment as key factors enabling religious tolerance. Religious tolerance as a fundamental value of Albanian society constitutes an element of unity for citizens of different religious backgrounds in the country.

While traditional religions are seen as part of a shared culture, they are not seen as insurmountable obstacles to mutual understanding and cooperation between religious divisions. These values of Albanians have not come as a requirement of the time but are the result of a particular history and culture. So, we may say that religious harmony in Albania lies between myth and truth. Both of these unite in the geo-political position of Albania in the Balkans and in Europe in particular. This position has served as an element of union between different religions, but also as an element of separation between east and west, between Catholicism and Muslimism. The relationship between the recognition and acceptance of a particular religion is still an unknown relation. As long as this report has not created conflicts, it remains intact and unexplored.

References

Giovanni Cimbalo, Pluralismo confessionale e comunità religiose in Albania, Bononia University Press, ISBN 978-88-7395-762-1, Elected Statute of Lushnja, 8-12-1922.

Ministero per i beni e le attivita culturali, Direzione generale per gli archivi, *L'unione fra l'Albania e l'Italia*, a cura di S. Trani, Roma, Pubblicazioni degli Archividi Stato, Strumenti CLXXIII, 2007,

Constitution of the Republic of Albania, Adopted by law no. 8417, dated 21.10.1998 of the People's Assembly. Approved by referendum on

---

[28] Study on Religious Tolerance in Albania, UNDP, Programi I Kombeve te Bashkuara per Zhvillim, 2018, https://www.undp.org/content/dam/albania/docs/religious%20tolerance%20albania.pdf

22.11.1998 Declared by Decree no. 2260, dated 28.11.1998 of the President of the Republic, Rexhep Mejdani Amended by Law no. 9675, dated 13.1.2007 Amended by Law no. 9904, dated 21.4.2008 Amended by Law no. 88/2012, dated 18.09.2012 Amended by Law no. 137/2015, dated 17.12.2015 Amended by Law no. 76/2016, dated 22.07.2016

Legal Statute of Religious Communities, http://licodu.cois.it/415/view.

Basic Statute of the Kingdom of Albania, 01.12.1928

Decree no. 241 dated 06.12.1946 "On the establishment of political parties", Official Gazette no. 115, dated 23.12.1946.

Civil Code of the Republic of Albania, approved by law. 7850 dated 29.07.1994

Law No.10 140, dated 15.5.2009 on the financing from the state budget of the religious communities, which have signed agreements with the council of ministers

Shyqyri Hysi, Myslimanizmi në Shqipëri *1945–1950*, Tirane, Mesonjtorja e pare, 2000.

Law nr. 5506 dated 28.12.1976 The Constitution of the Popular Socialist Republic of Albania.

Sokol Paja, PhD candidate , Universiteti Tiranes, Departamenti I Gazetarisë dhe Komunikimit, Book of Abstracts, First International Conference on Communication and MediaStudies, 29–30 May 2015, Kolegjia AAB, Pristina, Kosovo.

Study on Religious Tolerance in Albania, UNDP, Programi I Kombeve te Bashkuara per Zhvillim, 2018, https://www.undp.org/content/dam/albania /docs/religious%20tolerance%20albania.pdf

6

Tolerance and Peace Through the Portuguese Parliament Action

H.E. Fernando Negrão

Vice President of the Portuguese Parliament

Portugal is a small country, but one with great historic achievements and conquests. It was the first country in Europe to abolish the death penalty. Portuguese people are, by genetic definition, resilient. Citizens with this DNA are living proof of what John F. Kennedy had long been trying to teach: "If we cannot now end our differences, at least we can help make the world safe for diversity."

The Assembly of the Republic is the most important institution in the Portuguese political system, where legislative power is exercised, an open house that scrutinizes all the actions of the highest executive body in the Portuguese political system, and also empowers the youngest to be future leaders.

The Parliament has developed several instruments that enable tolerance and peace to take place, not only through the work of the standing committees but also through diplomatic interventions.

Under the complete respect of the principle of separation of powers, the Portuguese Parliament develops an outstanding role in the democracy and the satisfaction of the basic needs of the people living in this territory, maintaining tolerance and peace.

Keywords: tolerance and peace, Portuguese Parliament, peace-keeping, Portuguese DNA.

Tolerance and Peace Through the Portuguese Parliament Action

It is with enormous honor and admiration for the role of the Global Council for Tolerance and Peace across the world, in its President, H. E. Ahmed

Aljarwan, that I give my testimony, as Vice President of the Portuguese Parliament to the contribution of this Institution in these matters.

The 21st Century reveals long and diverse experiences throughout a multicultural world developing at different paces.

As Konrad Adenauer said, "We all live under the same sky, but we don't all have the same horizon."

Compared to others, Portugal is a small country but one with great historic achievements and conquests. It showed its universal and globalizing genesis early on, facilitating a meeting of different peoples and cultures, and sharing with its peers, the maritime knowledge it ventured to discover.

Portugal is a young democracy, achieved through the use of flowers, instead of the brutality of rifles and their ammunition.

It was the first country in Europe to abolish the death penalty, thereby providing compelling evidence of the humanism and pacifism inherent in the people of this great nation.

Doing justice to their fearless spirit, the Portuguese regularly leave their homeland to settle all over the world, often facing extreme difficulty. However, thanks to their tolerant, peaceful, and friendly spirit, they always manage to be effortlessly integrated into their host communities, ultimately being regarded as "children of the land."

Portuguese people are, by genetic definition, resilient. Despite the recent economic crisis that the country went through, all demonstrations of dissatisfaction and of the difficulties experienced were peaceful. The world did not witness demonstrations resulting in serious injuries or other significant consequences. This is the Portuguese DNA.

Citizens with this DNA are living proof of what John F. Kennedy had long been trying to teach: "If we cannot now end our differences, at least we can help make the world safe for diversity."

In an essay of the Francisco Manuel dos Santos Foundation, Jorge Fernandes states that the Assembly of the Republic is the most important institution in the Portuguese political system.

It is right here, in the Portuguese Parliament, that legislative power is exercised. This body is made up of women and men representing the citizens who elected them to be the active voice of their will, as reflected in the electoral program put to the vote.

In this sense, the Portuguese Parliament is an open house, available to welcome all those who wish to present their ideas. This house of democracy set up a program specially aimed at young people who wish to experience the role of members of parliament: whether it be by hosting groups of young people who put forward and discuss their ideas at the Parliament or by taking the Parliament and this experience to local schools with the help of Members of Parliament belonging to all political factions within their constituencies.

These experiences strengthen young people, empower them, and show them that they can find solutions to most of the issues that need solving in the daily life of governing a nation by presenting and debating ideas while maintaining peace and security in its territory.

Another significant milestone that distinguishes the Portuguese Parliament is the Human Rights Prize, created on the occasion of the 50th anniversary of the Universal Declaration of Human Rights, and which established the annual award of a monetary amount to reward the work of non-governmental organizations, prominent individuals, or other outstanding acts in this field.

It should also be mentioned that the world has witnessed a massive migration phenomenon at the beginning of this century, as a result of the wars still going on today, forcing populations to flee their countries to save their lives.

Europe has been the quintessential destination of this migration cycle, being under an enormous amount of pressure. Even at an earlier stage, the Portuguese Parliament and the legislation it produced have allowed Portugal to have been internationally recognized on several occasions as one of the best and most welcoming countries in the world. Last year, the United Nations distinguished the Portuguese migration services.

Parliamentary Groups of Friendship are a prominent figure for peace and tolerance also established by the Portuguese Parliament. They are set up between parliamentarians from different countries who may or may not have similar interests. However, they translate into a means of defining strategies for dealing with migration issues and the inclusion of other affairs, cultural differences, various legislative procedures, and cooperation in the field of membership of international organizations, among other advantages.

The first standing committee at the Portuguese Parliament, the Committee on Constitutional Affairs, Rights, Freedoms, and Guarantees, specializes in matters of tolerance and peace. They are responsible for matters including human rights, justice and correctional affairs, electoral law, migration and asylum issues, cross-cultural dialogue, border control, equality and non-discrimination, the fight against trafficking in human beings and domestic violence, as well as the protection of children and young persons at risk and of the elderly.

In this regard, it should be noted that Portugal was the first Member State of the European Union to ratify the Istanbul Convention: Council of

Europe Convention on preventing and combating violence against women and domestic violence. This Convention, which entered into force on 1 January 2014, resulted in the production of advanced legislation by the Portuguese Parliament, leading to fruitful actions in Portuguese society.

Nevertheless, maintaining the peace among the people of a nation requires meeting their basic needs, such as the supply of energy, water, housing; matters for which other standing committees are responsible (this distribution may vary in each legislature following the legislative elections preceding them).

In addressing these issues, with goals that can be achieved by different policy options, as is often also the case with environmental problems, public interest outweighs disagreements, and Portuguese parliamentarians frequently use a legal instrument which enables them to make recommendations to the Portuguese executive in order to ensure that solutions are found.

This was the case in the 13th legislature when all political forces came together and endorsed a resolution calling on the Government to draw up a mapping of housing needs in the country, as well as to examine solutions to address these shortcomings.

This is why, on top of the primary role of legislating, the Portuguese Parliament is "the most important institution in the Portuguese political system", which, in addition to recommending actions to the Government, has the power to scrutinize these and all other actions of the highest executive body in the Portuguese political system.

This scrutiny takes place in fortnightly debates, in which all members of the Government go to the house of democracy to "report" to the parliamentarians elected by the Portuguese people, but also in emergency debates requested by various political forces, as well as through parliamentary consideration of the diplomas drawn up by the executive body while exercising its power.

At the initiative of individual Members of the party groups to which they belong or even of the standing committees of the Assembly of the Republic, the Portuguese Parliament also has the power to hold field visits to assess situations to be addressed, such as, for example, the poor living conditions of a community of second- and third-generation Cape Verdean immigrants residing in Portugal, and to use the instruments provided for by the Constitution of the Portuguese Republic to act or arrange for the competent authority to do so.

All these means of action at the disposal of the Portuguese Parliament empower this institution as a significant and indispensable instrument for peace and tolerance in Portugal, as well as for Portugal's relations with its peers, bringing it closer to the population and opening it to their contributions, through the scheduling of hearings with Members of Parliament, parliamentary groups, or committees.

Always making use of its open and cooperative spirit, the Portuguese Parliament welcomes, on an institutional basis, delegations of other countries seeking information to compare existing legislation, practices, and results. Besides, there are exchanges of officials and staff with Parliaments in other countries, usually PALOPs — Portuguese-speaking African Countries, not only to share the knowledge and experience in constructing the Portuguese legal order which has made it possible to achieve this state of tolerance and peace in the country but, like all things in life, to allow for the acquisition of new knowledge, concepts, and also very productive ideas.

For all citizens to have access to parliamentary life, there is also an openly broadcast channel, ARTV, broadcasting the debates of the legislature, interviews with Members of Parliament and showing the actual Parliamentary area so that all citizens feel involved in democracy and political life in an inclusive manner. Political life in Portugal stands for tolerance and peace in the world as well: when we witness, for example, votes of condemnation, endorsed by all political forces, of the massacres of innocent people reported on all television channels.

The good diplomatic relations maintained by the Portuguese Parliament are also essential and indispensable nowadays for tolerance and peace throughout the world, thus keeping its historical tradition of external diplomacy.

The media operate today as an actual "particle accelerator" and the Portuguese Parliament, aware of this reality, does not overlook it and promotes information events on the infamous *fake news*. The role of the media is examined and scrutinized in one of the Parliament's standing committees, and all parliamentarians are aware of the harmful consequences of false information and of potential ways of recognizing it, as it can result in intolerant reactions that can disrupt the peace created in the Portuguese territory.

The principle of separation of powers applies in Portugal: legislative, executive, and judicial power. This separation is essential for peace-keeping, but the role of the Parliament is crucial also in this field since it legislates for the judicial power to implement the laws it produces.

Besides, even if doubts sometimes arise as to whether the judiciary is sexist, and whether there is gender stereotyping in judicial decisions, as reflected in Nelson Tavares's master's dissertation at the University of Coimbra, the law is increasingly equalitarian and has shown considerable development since 1988. Nonetheless, we must not forget the preconceived beliefs that could call into question the very impartiality of the judge, and that the law is responsible for paving the way for the necessary change in mind-sets leading to the equality required nowadays.

This is the only way for the Portuguese Parliament to help to ensure that tolerance and peace continue to prevail in our country as well, and, through some of its already mentioned instruments, share this path with its peers so that tolerance and peace are an achievement of the world and for the world.

References

"O Parlamento Português", de *Jorge Fernandes,* Fundação Francisco Manuel dos Santos

"Justiça machista? Uma análise sobre o estereótipo de género no seio das decisões judiciais", de *Nelson Filipe Correia Tavares,* Dissertação no âmbito do Mestrado em Ciências Jurídico-Forenses, Faculdade de Direito da Universidade de Coimbra (janeiro 2019).

7

Interreligious Dialogue and Its Contribution to International Security

Prof. Ferdinand Gjana

University College Bedër

Abstract

Over the past two decades, global security has been quite unstable, facing many new challenges. Since 11 September 2011, when hijacked aircrafts hit the towers in New York, the main rising threat of the international security has been terrorism. Deriving mainly from religious radicalism, it has caused many casualties in many countries around the world — without making any exceptions between east, west, south, or north. Many governments and key international organizations, such as the UN, NATO, and EU, have included security countering as a priority for preventing terrorism and violent extremism. These developed strategies can be classified into two categories according to the phase they are implemented: on one hand, there are preventive strategies and, on the other hand, there are countering strategies. Preventive activities such as education, inclusive development, interethnic, and interreligious dialogue can be labeled as a "soft way" and activities of countering terrorism, which mainly use military and policy interventions, can be classified as "hard ways." Preventive strategies, because there are many elements, need lot of time and efforts; in comparison to the "hard ways" of countering terrorism, they are more effective and less costly. These strategies comprise the inclusive development and promotion of tolerance and respect for diversity.

As terrorism during previous decades has been mainly motivated by religious ideologies, this makes the interreligious dialogue a very crucial action to prevent societies, religious group, and individuals from the manipulation and becoming part of the vicious cycle of terrorist activities. Many governmental and non-governmental organizations have been very active in the areas of promoting interfaith dialogue in different parts of the world. Religious coexistence and peaceful living has always been very important, but, at such a time, this can be considered of a vital significance. This coexistence besides the important role to keep social and political sustainability in the country also has a contribution to international security. This article analyzes the interreligious dialogue and its contribution to international security by giving several examples and especially the case of Albania.

Keywords: international   security,   religion,   interreligious   dialogue, terrorism.

7.1 Traditional Meaning vs. the New Environment of International Security

When you look at the meaning of security, the change that it has undergone is obvious. In the past, the answer to "which is the safest country in the world?" was easier to find in comparison with the new environment. When you analyze some other dimensions of the traditional international security and the new environment of it, you realize easily how actors have changed: in a more traditional meaning, governments, militaries, and military alliances were the main actors of security — but now, the number of these actors has increased by also including corporates, media, civil society, terrorist groups, etc. Upon analyzing threats in the past, it becomes clear that this was an external power — but now, you are surrounded everywhere by different risks that can spoil your peace. In this new environment, security has become broader and broader. This transformation is best described as a broadening and a deepening of the security agenda in Paris 2001.[1] "Broadening" the security agenda implies the inclusion of non-military threats such as terrorism, as well as security challenges, such as environmental scarcity, pandemics, or mass refugee movements. "Deepening" the security agenda

---

[1]Le Gloannec, Anne-Marie; Irondelle, Bastien, and Cadier, David (Ed): New and Evolving Trends in InternationalSecurity, The Transatlantic Relationship and the future Global Governance, Working Paper 13, April 2013. Available from: https://spire.sciencespo.fr/hdl: /2441/2tb654p5g79rg9aejj1ah6job0/resources/tw-wp-13.pdf (accessed 22 March 2020)

means considering referent-objects other than the state, such as individuals, social groups, or planet Earth. These two dynamics are interlinked since addressing non-military threats and challenges often entails moving beyond states as referent-objects.[2]

7.2 International Security and the Scary Trends of Terrorism

Throughout the last decade, one of the main threats to the new security environment has been terrorism and violent extremism. There are different statistics on terrorism which should be taken very cautiously for conceptual and methodological reasons. In general terms, however, it appears that terrorist violence has been on the rise and has become more dangerous,[3] to the extent that, today, it is commonplace to say that terrorism is a major threat to international security. When speaking in general terms, the causes of terrorism might be many, such as ethno-nationalism, alienation, discrimination, socio-economic status, and political grievance, but the recently risen terrorism reports have derived from religion. Over the 15 years, South Asia, Egypt, Turkey, Germany, France, Spain, and many other countries and regions have been under a high attack of terrorist groups by causing a huge amount of deaths and other costs. According to Global Terrorism Index (GTI; 2017), terrorist attacks against civilians have increased by 17% from 2015 to 2016 at a global level and the primary targets of them have been private citizens and property.[4] The 2019 Global Terrorist Index underlines that although the intensity of terrorism has diminished, its breadth has not;103 countries recorded at least one terrorist incident in 2018, and 71 countries suffered at least one fatality in that same year. This is the second worst year on record for the number of countries suffering at least one death, which highlights the need for continued assertive international action to combat terrorism.[5] The most dangerous terrorist groups, which have been responsible for 59% of all deaths

---

[2]Ibid.

[3]Lutz, James; Lutz Brenda Global Terrorism, Third Edition, Routledge, 2013.

[4]Institute for Economics & Peace. Global Terrorism Index 2017: Measuring the Impact of Terrorism, Sydney, November 2017. Available from: http://visionofhumanity.org/reports (accessed 21March 2020).

[5]Institute for Economics & Peace. Global Terrorism Index 2019: Measuring the Impact of Terrorism, Sydney, November 2019. Available from: http://visionofhumanity.org/reports (accessed 21March 2020).

in 2016, are ISIL, Boko Haram, al-Qaida, and the Taliban. By killing 9132 people in 2016 with the majority of these deaths occurring in Iraq, with a 50% increase in deaths from its previous peak in 2015, ISIL has been the deadliest group since 2016.[6] Terrorist attacks in 15 countries have been undertaken by ISIL only in 2016, which have been drastically rising in comparison with previous years. The other three deadliest terrorist groups, Boko Haram, al-Qaida, and the Taliban, have been also responsible for a very high number of deaths in many attacks in different countries. Ending these terrorist groups has been a great challenge for all governments and international organizations through a much-expanded strategy starting with means of prevention and countering in many ways.

When analyzing the above-mentioned terrorist groups, it is obvious that all of them have a common ideological background, which is supported by religious radicalism approaches ideologies. In addition, these groups' leaders or any affiliated individuals to these groups have continuously announced themselves as Muslims believers that aim to reach their objectives through this violent and insane way. According to GTI(2017), the main factor of today's conflict seems to be religion and religious difference, and againthe GTI (2017) further affirms that terrorist acts are committed in 67 countries out of 167 in the world today, and 51% of them is on religious ground.[7]

7.3 Interreligious Dialogue as an Effective Way of Preventing and Countering Terrorism and Violent Extremism

By having a glance at policy papers and strategies dealing with countering terrorism and prevention of violent extremism from different international organizations such UN, OSCE, and NATO, you can easily realize that there are so many approaches integrated into these strategies. They include hard ways, which includes police and military force — and at the other side,

---

[6] Institute for Economics & Peace. Global Terrorism Index 2017: Measuring the Impact of Terrorism, Sydney, November 2017. Available from: http://visionofhumanity.org/reports (accessed 21 March 2020).

[7] Institute for Economics & Peace. Global Terrorism Index 2017: Measuring the Impact of Terrorism, Sydney, November 2017. Available from: http://visionofhumanity.org/reports (accessed 21 March 2020).

there are many soft ways indicated such as education, inclusive development, and improving the rule of law and human rights. One of the most critical of this particular category is that of increasing interreligious dialogue.[8]

"There will be no peace between the civilizations without a peace between the religions! And there will be no peace between the religions without a dialogue between the religions."[9]This is one of the main statements of Hans Küng, who claims and expects interreligious dialogue to advance peace and promote security on earth.[10] Countering and preventing religious terrorism is in the center the indoctrinated and sometimes misinformed and misused individuals. These people are almost impossible to have a successful treatment process just with use of hard ways, which may be undertaken by government institutions such as military or police forces. It is very indispensable with a high impact, and the role might have the religious institutions and leader over this indoctrinated/ideologically "infected" group to facilitate the process and speed up the solutions. It is usually easier, more effective, and less costly[11]to prevent such a phenomenon in social, political, and economic dimensions than to counter it after many damages and problems have been risen. When you see the statistics on terror acts, conflicts, and wars, they further prove that the world has to unite to enhance religious tolerance, interfaith dialogues, and mutual understandings. For this reason, there have been many actors — such as governments, intergovernmental organizations, international organizations, civil society, faith-based organizations, and NGOs, engaged in raising the awareness of the importance of religious tolerance and improving the interreligious dialogue. Besides the important actions undertaken by many international actors, a substantial number of states have been active in countering the phenomenon of terrorism, violent extremism, and religious radicalism to ease this serious worldwide problem.

---

[8]Preventing Violent Extremism through promoting inclusive development, tolerance, and respect for diversity(UNDP) Available from: https://www.undp.org/content/undp/en/home/li brarypage/democratic-governance/conflictprevention/discussion-paper---preventing-violent -extremism-through-inclusiv.html

[9]Küng, Hans: A Global Ethic for Global Politics and Economics. Trans. John Bowden. p.92, New York: Oxford University Press, 1998.

[10]Ibid.

[11]Many analyses argue the high economical losses caused by terrorist attacks but and also the high cost of countering it by hard ways, i.e., military or police interventions.

7.4 Examples of International Initiatives in Promoting the Interreligious Dialogue

As the concept and the dynamics of international security drastically changes, preserving global security and peace building is one very important issue of the agenda of all political actors such as states, international organization, NGOs, civil society, media, and even corporates. Over the last two decades, all these actors have been in action to contribute to a more secure world. Organizations like the UN, OSCE, NATO, and many others have integrated in their strategic planning many parts related with new era security challenges. The same has occurred with several governments in different parts of the world or NGOs and other international institutions. The majority of these activities have also included many initiatives in the field of interfaith dialogue by emphasizing that this can have a critical role in preserving security and promoting peace.

7.5 Active Role of UN in Interreligious and Intercultural Dialogue

The UN has been playing a very active and important role to promote interreligious dialogue by preparing and adapting many resolutions.[12] The UN has played a very active role by stressing continuously that interfaith initiatives can ensure rich cultural diversity to make the world more secure and to increase the culture of peace.[13] One important initiative of the UN is the adaptation by the General Assembly of draft resolution A/61/L.60 entitled "High-Level Dialogue on Interreligious and Intercultural Understanding and Cooperation for Peace," on 25 May 2007.[14] In this resolution, the UN has considered as a great need the Interreligious and Intercultural Understanding and Cooperation for Peace and this has attracted increased attention amongst governments, UN agencies, religious communities, spiritual movements, civil society, and humanists at the beginning of the 21st century. The objectives of

---

[12] Available from: https://www.un.org/press/en/2019/ga12226.doc.htm (accessed 21 March 2020).

[13] Available from: https://www.un.org/press/en/2008/ga10782.doc.htm (accessed 21 March 2020).

[14] Concept Note Informal Interactive Hearing with Civil Society 2007 High-Level Dialogue of the General Assembly on Interreligious and Intercultural Understanding and Cooperation for Peace, Available from: https://www.un.org/press/en/2007/ga10630.doc.htm (accessed 22 March 2020).

the "High-Level Dialogue" have been to strengthen efforts of interreligious and intercultural understanding and cooperation by engaging a variety of actors and constituencies, especially in government, civil society, and the United Nations system.[15] These three parties have also been at the core of the Tripartite Forum on Interfaith Cooperation for Peace formed after the 2005 Conference on Interfaith Cooperation for Peace. The High-Level Dialogue further has been continuously seeking to promote a culture of peace and dialogue among civilizations; advance multi-stakeholder coalitions, including the private sector on related issues, further strengthen the Alliance of Civilizations initiative and translate shared values into action in order to achieve sustainable peace in the 21st century.[16]

7.6 The Active Role of Vatican in Interreligious Dialogue

The Second Vatican Council was the first ecumenical council in the history of the Church to give serious consideration to the Church's relationship to the followers of other religions and to advocate interreligious dialogue as an integral dimension of her mission. As is generally recognized, the Declaration Nostra Aetate, although one of the shortest of all the documents of the Second Vatican Council, is one which has had a considerable impact on the life of the Church. This document has served to open the door for Catholics into the world of interreligious dialogue and since then had gone through several drafts and its relationship with other documents of the Council had varied.[17] The leadership of Vatican has been engaged with the interreligious dialogue according to the level of necessity they have considered to be in different periods. Some Popes have been obviously much more active in interreligious initiatives and activities than others. Pope Francis can be clearly observed to have taken the issue of dialogue as one of the top priorities during his leadership.[18] Almost in every trip, Pope Francis seems to have had an interreligious or ecumenical moment. One very important meeting in this

---

[15] Concept Note Informal Interactive Hearing with Civil Society 2007 High-Level Dialogue of the General Assembly on Interreligious and Intercultural Understanding and Cooperation for Peace, Available from: https://www.un.org/press/en/2007/ga10630.doc.htm (accessed 22 March 2020).

[16] Ibid.

[17] Fitzgerald, Michael L.: Nostra Aetate, a Key to Interreligious Dialogue, Gregorianum, Vol. 87, No. 4 (2006), pp. 699–713, GBPress-Gregorian Biblical Press.

[18] Youth, interfaith dialogue, and peace dominate Pope's foreign trips in 2019, Available from: https://cruxnow.com/vatican/2019/12/youth-interfaith-dialogue-and-peace-dominate-p opes-foreign-trips-in-2019/ (accessed 21 March 2020).

context came in February 2019 when he visited the United Arab Emirates and signed a declaration of the Human Fraternity Document for World Peace and Living Together with the Grand Imam of Cairo's Al-Azhar University.[19]

7.7 UAE Initiatives in Promoting Religious Tolerance and Dialogue

During recent years, one very active country in contributing to international security and peace through promoting interreligious dialogue has been the United Arab Emirates. The government of this country has undertaken many steps at the domestic and international levels to support the peace and dialogue culture serving a more secure world. One of the very important initiatives of UAE was proclaiming officially 2019 as the Year of Tolerance,[20] aiming to strengthen the nation's role of encouraging stability and prosperity in the region and the promotion of culture of tolerance, dialogue, and peace. The UAE has hosted many important dialogue summits and events at home and abroad, calling for religious tolerance and a warning that countries must stand up to all forms of extremism.[21] One of the most important activities was in February 2019, the organization of a Global Conference at Abu Dhabi, hosting the Pope Francis and the Grand Imam of Al-Azhar Ahmet el-Tayeb. During this event, Pope Francis and Grand Imam Ahmet el-Tayeb signed the Human Fraternity Document for World Peace and Living Together, which can be considered one of the most important written acts in history which formalizes the dialogue between Christians and Muslims.

7.8 Interfaith Dialogue in Albania: A Multi-Dimensional Example of Coexistence

The concepts of the interfaith dialogue, religious tolerance, and coexistence in Albania are something that everybody is familiar with. During the communist regime, there was no religious freedom in Albania; furthermore, all religious cults were closed, most of them demolished, and clerics were persecuted and imprisoned. With the fall of communism, besides other freedoms, the

---

[19] Available from: https://www.vaticannews.va/en/pope/news/2019-02/pope-francis-uae-d eclaration-with-al-azhar-grand-imam.html (accessed 22 March 2020).

[20] Available from:https://www.theyearoftolerance.ae/en (accessed 21 March 2020).

[21] Available from: https://www.thenational.ae/uae/government/sheikh-abdullah-tells-of-nee d-for-greater-interfaith-dialogue-1.824123 (accessed 21 March 2020).

freedom of faith came into life and religions started their activities. Actually, there are five officially recognized religious communities in Albania: Muslim Community of Albania, Orthodox Church, Catholic Church, Bektashi Community, and Evangelical Brotherhood. Besides these religious groups, there is freedom for religion that permits also other religions for their activists such as Jehovah's Witness, Baha'i, Shia, etc., but they are very few in society. According the census of 2011, 56.7% of the population was declared to belong to the Muslim faith, 10% Catholics, 6.7% Orthodox, and 2.1% Bektashis.[22]

7.9 Historical and Social Dimensions of Religious Tolerance and Coexistence in Albania

Religious tolerance is not just a recent post-communist phenomenon in Albanian society. When you go through history, there are several good examples that explain this positive atmosphere of living together. This is shown also by the role of the religious leaders in the independence process of Albania, who played a very crucial mission building the newborn state of Albania. The Muslim religious leaders, such as H. Vehbi Dibra, were the figure who approved the flag during the independence. He and many other outstanding personalities played a decisive role in building the national institutions of the independent Albania. Another example mentioned in history is the saving of the Jews from deportation to the concentration camps during World War II. The Albanian rescue is unique because of its scale, including not just Jews within their borders but also every Jew who sought refuge there during the Holocaust. When analyzing the interreligious dialogue in Albania, it is indispensable to talk about its situation in grassroots, in society. After a quick glance, it is easily noticed that all religions live together —no separation of the cities and no division in quartiers or in any other form of the population. There are different regions and cities, which might have a majority of one religion, but all are living together without any exclusion. You can see this social unity of religion in every Albanian family where you can find too many mixed marriages and many relatives having different

---

[22] In 2011, the Albanian government conducted the Census process and one of the questions of this census was also that "to what religion you belong?"INSTAT/Census 2011.Available from: http://www.instat.gov.al/al/temat/censet/censusi-i-popullsis%C3%AB-dhe-banesave/#tab1 (accessed 21 March 2020).

religions from each other. According to Emory S. Bogardus, theoretically analyzed, if there are many interreligious marriages, if the interreligious close friendship in schools and working place is widespread, if the people live altogether in a good atmosphere in a city, or if a quartier or an apartment nevertheless has the religion identity, it means that there is no religious social distance in such societies.[23] There may be several reasons for this social reality, but the most important one is of having the same ethnic origin although it is a multi-religious population and this has united Albanians and made them live together and survive, despite different external pressures in history.

7.10 The Role of Religious Institutions and Their Leadership

The role of religious institutions has been one of the most important and influential in cultivating and keeping the interreligious coexistence in Albania. There is high respect and a credibility shown by society to religious institutions and their leadership. During a speech in 1995, the leader of MCA was declaring, "somebody was asking me how many Muslims live in Albania? In addition, my answer was: It does not matter. We are not going to fight with each other".[24] This situation is more critical of all the messages delivered by religious institutions. By analyzing the preaches, sermons, and other activities, you can easily find too many examples of when they have been promoting common values and mutual respect. The language used by all religious leaders and other representatives, predominantly, has been that of love and tolerance. The impact of religious communities on society can be considered to be at a high level and, in many cases, they have been very active in finding solutions to various problems in society. There is also another assumption regarding the influence of the religious representatives in Albanian society saying that they are not so credible — but one thing for certain is that if any religious group or community, even if it is small, causes a very turmoil situation, they behave contrary to the social peace principles. The main reason for this interreligious coexistence in Albania

---

[23]Bogardus, Emory S.: "A Social Distance Scale", Sociology and Social Research 17 (1933): 265–271.

[24]The speech of Head of Muslim Community of Albania, H. Sabri Koçi, delivered in a Mosque in the 1990s.

can be encountered to be the moderate teaching approach of the religions followed by their respective representative institutions.

So, it is very important to have religious communities, their leaders, and representatives preach their religion by promoting, first and foremost, love and mutual respect, and especially working to keep their communities away from harmful behaviors and violence. Although Albania has historically been considered a good environment, interreligious relationships should not be taken for granted. For these reasons, with the encouragement of the respective religious institutions, the three theological schools of higher education — which are supposed to prepare the future needed religious human resources — are organizing together in a continuous manner joint program, conferences, symposium, workshops, etc., to get to know each other better and to prevent any potential misunderstandings that might spoil the respectful and lovely atmosphere of society.

7.11 State, Politics, and Religions: A Balanced Equilibrium

The Republic of Albania Constitution starts by saying that, "We, the people of Albania, proud and aware of our history, with responsibility for the future, and with faith in God and/or other universal values, with determination to build a social and democratic state based on the rule of law, and to guarantee the fundamental human rights and freedoms, with a spirit of religious coexistence and tolerance, with a pledge to protect human dignity and personhood, as well as for the prosperity of the whole nation, for peace, well-being, culture and social solidarity, with the centuries-old aspiration of the Albanian people for national identity and unity, with a deep conviction that justice, peace, harmony and cooperation between nations are among the highest values of humanity."[25]

In addition, it continues on by stating that:

1) "In the Republic of Albania there is no official religion."
2) "The state is neutral on questions of belief and conscience and guarantees the freedom of their expression in public life."
3) "The state recognizes the equality of religious communities."

---

[25]Constitution of the Republic of Albania, text approved by referendum on 22 November 1998 and amended on 13 January 2007, translated under the auspices of OSCE-Albania.

4) "The state and the religious communities mutually respect the independence of one another and work together for the good of each and all."

5) "Relations between the state and religious communities are regulated based on agreements entered into between their representatives and the Council of Ministers. The Assembly ratifies these agreements."

6) "Religious communities are juridical persons. They have independence in the administration of their properties according to their principles, rules and canons, to the extent that interests of third parties are not infringed."[26]

These articles of the constitution of Albania explain very well the position of state and the religious communities within the society. The state institutions and the religions have been too careful to behave according to the constitution and the results seem too positive enough in having a good equilibrium in this relationship. The state has signed agreements to regulate the terms of the relationship with five religious communities and the assembly has ratified these agreements.[27] The use of religions for different interests by political actors can be very dangerous and is not a healthy approach to be followed in a multi-religious and democratic society. Albanian politics have been careful and sensitive in subjects related to religions and religious communities. Among the main political parties, there is no one putting on the agenda, specifically the religious-oriented subjects or having a closer intention to work with a specific religious group.

There have been continuously supportive discourses to religious communities from the state leaders and the outstanding political leaders, keeping in mind the principle of the laicite that the constitution has.

7.12 The Interreligious Coexistence in Albania as Part of Its Public Diplomacy

Albania is a multi-religious society and a Muslim-majority country on the European continent. This fact would be an important characteristic mentioned

---

[26]Ibid.

[27]There are agreements of Republic of Albania with the five religious communities: Catholic Church, Muslim Community, Orthodox Autocephalous Church, The Holly Seat of the World Bektashi, and Evangelical Brotherhood. (Law No. 9365, Date 31.3.2005; Law No.10056, Date 22.1.2009; Law No.10057, Date 22.1.2009; Law No.10 058, Date 22.1.2009; Law No.10394, Date 10.3.2011).

by several political and religious world leaders on many occasions. When US President George W. Bush visited Albania in 2007, he, prior to his arrival, declared, "I will visit a Muslim-majority country coexisting in peace with all religious groups."[28] Pope Francis, in 2014, would mention several times prior and after his visit the characteristics of religious tolerance, interreligious coexistence, and dialogue by emphasizing that this should be kept unspoiled because it is a very important value of Albanian society in the age we are living in. Albanian governments, through the years, have tried to promote this valuable asset to the international arena by showing that if there are problems in multi-religious societies deriving from the religion, a solution can be found to overcome it as there is such an example of a country living in peace and mutual respect. The state representatives of high levels[29] have continuously presented and promoted in many international summits and meetings, especially in those dealing with countering religious radicalism and violent extremism, as a good example of interreligious coexistence in Albania. In many international initiatives and projects, although a small country with limited resources and capacities, the Albanian government has tried to play an active and crucial role in finding solutions to the problems deriving from religions around the world. The interreligious coexistence of Albania has been mentioned and pointed out as a positive example to be taken into consideration finding a solution in other countries and societies where needed.

7.13 Conclusion

The threats of international security have changed, and it seems that they will have very different and complex discourse in the future. To maintain a secure world, the tools of countering these threats should be varied and adapted according to the environment and its needs. It is obvious that classical ways of maintaining international security are insufficient toward very intelligent and integrated menaces which always are looking for the weak points to harm the countries, societies, and all humanity. One of these alternative ways to sustain security at all levels is by using increasing interreligious communication and dialogue. Good initiatives in this field have been undertaken by many national

---

[28] Available from: https://www.reuters.com/article/amp/idUSTRE7655J520110706 (accessed 21 March 2020).

[29] Presidents, Prime Ministers, Ministers of Foreign Affairs, etc.

and international actors and the results have proven the good efficacy and the high impact of these actions. Although a small country, Albania has tried to show a good example of how interreligious coexistence and dialogue can positively affect society.

By being a tangible case, Albania offers a positive model in interfaith harmony and coexistence to the multi-religious societies and this is, by itself, a significant way to contribute to international security. However, the question is this: can this model be exported and implemented in other countries and societies, Europe, or elsewhere? Apparently, the exportation and implementation of any social model to somewhere is not an easy process and sometimes might be quite impossible. To realize such a project, it is needed to enable all social dynamics and reactivate them in a new society. However, you can try to adapt that social model by splitting it in parts, even if you cannot implement the whole, you can do it partially. Second, according to the dynamics of each country and society, you can use good examples you have acquired to develop strategies and to come up with new models. Of course, such a process is not easy because this is not a corporate structure or an engineering project where there are some mathematical variables, including principles and values, which can be easily transferred. The relocation of the principles and values and their application in different societies requires a huge adaptation to the new social environment. Bearing in mind this information and the fact that the origins of religious radicalism driving into terrorism are very correlated with many social problems, the best way to solve this would be realized through a recipe of social sciences. In addition, within this recipe, the promotion of a model that contains tolerance and interfaith dialogue in the center would act as the main cure. In this way, we would achieve our goal and the solution would be more effective and sustainable.

References

Casey Lucius, *Religion and the National Security Strategy*, Journal of Church and State, Vol. 55, No.1 (Winter 2013), pp. 50–70, Oxford University Press.

Constitution of the Republic of Albania, Text approved by referendum on 22 November 1998 and amended on 13 January 2007, Translated under the auspices of OSCE-Albania.

Db Subedi And Bert Jenkins , *Preventing And Countering Violent Extremism: Engaging Peacebuilding And Development Actors*, Counter Terrorist

Trends and Analyses , Vol. 8, No. 10 (October 2016), pp. 13–19, International Centre for Political Violence and Terrorism Research

Edited by Anne-Marie Le Gloannec, BastienIrondelle and David Cadier; Contributors: NelliBabayan, David Cadier, BastienIrondelle, Anne-Marie Le Gloannec and Thomas Risse; *New and Evolving Trends in International Security*, The Transatlantic Relationship and the future Global Governance, Working Paper 13, April 2013.

Institute for Economics & Peace. Global Terrorism Index 2017: Measuring the Impact of Terrorism, Sydney, November 2017. Available from: http://visionofhumanity.org/reports

Institute for Economics & Peace. Global Terrorism Index 2019: Measuring the Impact of Terrorism, Sydney, November 2019. Available from: http://visionofhumanity.org/reports

James Lutz and Brenda Lutz, *Global Terrorism*, Third Edition, Routledge, (2013)

Hans Küng, *A Global Ethic for Global Politics and Economics,* Trans. John Bowden, New York: Oxford University Press, (1998).

Michael L. Fitzgerald, *Nostra Aetate, a Key to Interreligious Dialogue* : Gregorianum, Vol. 87, No. 4 (2006), pp. 699–713, GBPress- Gregorian Biblical Press.

NilaySaiya, *Religion, Democracy and Terrorism*, Perspectives on Terrorism, Vol. 9, No. 6 (December 2015), pp. 51–59, Terrorism Research Initiative.

Peter Admirand , *Dialogue in the Face of a Gun? Interfaith Dialogue and Limiting Mass Atrocities*, Soundings: An Interdisciplinary Journal , Vol. 99, No. 3 (2016), pp. 267–290 , Penn State University Press.

Resolution adopted by the General Assembly on 20 December 2006 [*without reference to a Main Committee (A/61/L.11/Rev.2 and Add.1)*]

Ron E. Hassner and Michael C. Horowitz, *Debating the Role of Religion in War*, International Security, Vol. 35, No. 1 (SUMMER 2010), pp. 201–208, The MIT Press

Thomas J. Badey, *The Role of Religion In International Terrorism*, Sociological Focus, Vol. 35, No.1 (February 2002), pp. 81–86, Taylor & Francis, Ltd.

Thomas Scheffler, Interreligious Dialogue and Peacebuilding, Die Friedens-Warte, Vol. 82, No. 2/3, Religion, Krieg und Frieden (2007), pp. 173–187, Berliner Wissenschafts-Verlag

ZeevMaoz and Errol A. Henderson (eds): *Scriptures, Shrines, Scapegoats, and World Politics*, University of Michigan Press. (2020)

8

Universities as Cosmopolitan Places for a Culture of Peace and Tolerance. The Case of the Ponto De Partida — Experiências Educativas

Cláudia Vaz, Ana Carolina Reis, João Ferreira Marini

University of Lisbon, Portugal

Abstract

The *Ponto de Partida — Experiências Educativas IN* | *The Starting Point — Educational Experiences IN* (INclusive, IN[non]formal and Inspiring), like other similar university projects, most likely resulted from something as simple as a *Kantian* impetus stemming from *goodwill and a sense of duty* — an informal conversation in a university hallway between a professor and an African student. Professor — "Is it true that African students at our university face greater difficulties and challenges than other students?" Student — "Oh Professor, you cannot imagine how much greater. In particular, students from Guinea-Bissau face difficulties you cannot even imagine." The professor may then respond with, "We must do something to change this situation." To that, the student replies, "We absolutely must!" The professor, eager to address this issue, states, "Let's schedule a meeting with the African students. I would like to hear what they have to say, get to know their experiences here and see what we can do from now onwards." Then finally replies — "Yes, let's do it!"

Guided by the assumption of *the dignity of the human person* (Kant 2010) from a cosmopolitan perspective (Appiah, 2006; Becker, 2016), in little over two years, this simple *act of goodwill* imbued with a sense *of duty* gave rise to a methodology for the development of the human potential in a multicultural context which aims to educate students to be the best version of themselves, regardless of whether in the job market, as political citizens or as human beings. Through this methodology, we have demonstrated that educational experiences in "cosmopolitan places" are transformative in the sense that they promote soft skills (such as empathy, creativity, problem-solving, critical thinking, teamwork, and communication) which are in themselves essential for building a culture of tolerance and peace.

Keywords: cosmopolitanism, ethics, development of human potential, *soft skills*, culture of peace and tolerance, starting Point — Educative Experiences IN.

8.1 Introduction

The objective of this chapter is to reflect on the role that universities in the 21st century play in training young people to build a culture of tolerance and peace. The proposal stems from the *Ponto de Partida — Experiências Educativas IN | Starting Point — Educational Experiences* (INclusive, IN[non]formal and INspiring), an educational project which aims to create educational and cosmopolitan environments that, on the one hand, stimulate the development of *soft skills* (such as creativity, emotional–empathetic intelligence, problem-solving, critical thinking, teamwork, big-picture thinking, and interpersonal communication) in a global and VUCA (volatile, uncertain, complex, and ambiguous) world and also foster the successful inclusion of international students.

In just over two years, from design to prototyping and execution, more than 40 *activity-experiences* were carried out in the classroom and in spaces outside the university (in other institutes and colleges at the University of Lisbon and in the streets and cosmopolitan spaces of Lisbon city), with groups of students from ISCSP, University of Lisbon, international universities,[1] and trainees from companies (collaborative knowledge transfer). The scientific areas were diverse.

Similarly, to Serres, "I would like to write narratives, songs, poems, a thousand enthusiastic texts encouraging all women and men to intervene in all public matters that are, and are not, pertinent to them" (2019, pg. 114), with the aim of creating a true culture of tolerance and peace. But I will not do it, not just yet (and certainly not alone). What I propose here is a collective reflection on a cosmopolitan pedagogic design adapted to 21st

---

[1]Cases from Warsaw Universities (Poland), Windesheim University of Applied Sciences (the Netherlands), and the College of engineering at Texas University, the latter represented by the company Brazil Cultural.

century universities, in line with a global agenda of building a culture of tolerance and peace. This too is the view of Ozdemir (2016), "To respond to the challenges of a globalized world, new visions and mindsets are needed. This can be done by 21st century universities."

To reflect on this, we begin with four assumptions.

1) We believe that *Starting Point's* activities are true cosmopolitan experiences and are places for personal transformation because they mirror the *Other*, the one that is different from myself and those who are closest to me.

2) We believe that these places, designated here as *cosmopolitan places*, facilitate essential learning, designated as *soft skills* that are crucial for our personal, social professional, and environmental success of millennials, centennials, and other generations of the 21st century.

3) We believe that the University, as an institute of higher learning, is responsible for producing high-level professionals ready to tackle market challenges in a global world and responsible for producing researchers and cultivating human and environmental knowledge. In addition, it also has the responsibility to incorporate and support cosmopolitan programs for the development of *soft skills* in order to guarantee human potential in this *Baumanian liquid modernity*, which characterizes today's world.

4) We believe that a humanist education with humanity, is the path to "constructing the *builders*" of a culture of tolerance and peace. During this process, it is interesting to understand the role of inspiring narratives and its play in forming positive beliefs and values. Although the urgency to reflect and contribute toward building a culture and a narrative of tolerance and peace are unfortunately not a new theme, thankfully, various initiatives, on a global as well as a national and local scale, have sought to answer this global appeal that has at its root the Kantian Idea of the *Dignity of the Human Person*.[2] There are also several contributions that indicate the importance of education curriculums focusing on peace that integrate activities which incite the development of emotional skills.[3]

---

[2]In this regard, see the works of Nussbaum (2003, 2012), Sen (1985, 2009), Reardon & Snauwaert (2011), Kant (2018, 2019), and Popkewitz (2009), among others.

[3]This is the case of Reardon (2001) in *Education for a Culture of Peace in a Gender Perspective*. In this author's perspective, skills of co-operation, communication skills, cultural skills, and conflict skills are some of the social skills essential for building common purposes.

As you will have the chance to see, our perspective is based on the originality of this systemic-multilevel design. For one, we have its hybridism, followed by a collaborative nature emphasizing its synergy. In conjunction, with a defined flexibility that can be farmable to any situation as a mechanism to which one uses and not necessarily as a final form.

This chapter is organized as three sections, resulting in a crossing of both objectives and assumptions presented, with the design of the non-formal, inclusive, and inspirational exercises that are Starting Point's cosmopolitan experiences.

Therefore, the initial "(One) Starting Point" is dedicated to this case study based on its context, process, and model. The third section — "Assumptions of the Model and Pillars of Education — An analogy," is committed to situate the model's assumptions within the scope of education for the 21st century. Finally, the fourth section — "The responsibility of Universities *to Nurture* a Culture of Tolerance and Peace," is organized around two questions: How can models like the SP-EEIC be an integral part of a more socially responsible educational culture? How can Universities contribute to a culture of peace and tolerance?

8.2 (One) "Starting Point" — An Educational Experience

Starting Point, as a university project of just two-years-old, aims to create cosmopolitan educational environments that trigger, on the one hand, the development of soft skills (empathy, creativity, emotional intelligence, problem solving, teamwork, big picture thinking, cognitive flexibility, and intercultural communication) in a global and VUCA world and, on the other hand, a successful integration of international students (with a special focus on Lusophone students). It is, therefore, a humanist project for the development of human potential.

Like so many other university projects, it resulted from something as simple as a Kantian impulse of feelings of goodwill and duty — an informal conversation in the hallway of a university between a Guinean professor and student. A conversation as simple as this:

- *Is it true that African students arrive at our university with greater difficulties compared to other students?*
- *Yes Professor, you can't imagine how many difficulties they have to overcome on a daily basis.*
- *We have to do something to change this situation.*

- *We really do!*
- *We will schedule a meeting with African students. I want to hear what they have to say, to know their experiences and see what we can do from there.*
- *Let's go!*

Understanding the statement that Starting Point was born from the intersection of feelings of goodwill and duty goes through the knowledge of the characters in the story — but is not that always the case? Is there a project, an idea, or even a business that can be truly understood without taking into account its protagonists? If so, this is not the case.

I am the professor of this story. I majored in Anthropology, the human science that is concerned with the understanding of behaviors and attitudes of all human beings. I do not know if being born in Africa, more specifically in Angola, will explain my anthropological interest in hybrid identities of children and young people of African origin (with a focus on those of Cape Verdean origin).[4]

Regardless of the reasons why I chose this master's degree, I studied the identities of children of Cape Verdean origin in school and for my Ph.D., the identities of young people from one of the most stigmatized migrant neighborhoods in Portugal — Bairro do Alto da Cova da Mora — fantastic experiences of learning and growth.

Therefore, I have always considered myself a lucky person because I had the opportunity to meet and relate to different people. I truly do not know of a better way to learn and become who we are as individuals than this.

Later, I had the opportunity to co-ordinate "Ser Mulher" (Being a Woman), in Portuguese — a project that aimed to 1) introduce the faces, voices, and narratives of young people and women who could, on the one hand, constitute positive references for other women (inspirations in women) and, on the other hand, stimulate the identity awareness of being a woman in a Portuguese-speaking country, and 2) to establish the link between the academy, policymakers, the media, and civil society (advocacy strategy). Following this project, I was invited by the President of ISCSP to represent this higher education institution on the thematic commission on Education, Higher Education, Science and Technology of the Community of Portuguese Speaking Countries (CPSP), in the role of Consultant Observer. One of the

---

[4]Until the 1990s, the Cape Verdean population represented the most numerous ethnic group in Portugal (they began to arrive massively in the 1970s, as a result of the decolonization policy).

topics that have systematically integrated the agenda is the mobility and integration of students from this specific community.

It was in this context that, one afternoon, I came across another protagonist of this episode — Maquilo Jamanca, who, at the time, was a pupil of mine finishing his master's in Strategy. He was born in Guinea (in Bissau, Bafatá) and came to Portugal at the age of 21, with the intention of continuing his studies in a Portuguese higher education university. Before entering university, he did his 12th year in languages and humanities at a school in Montijo. As you will observe, it was a meaningful experience for him:

Right on the first attempt to complete 12th grade here, I failed at the Portuguese language! I was always a good student in Portuguese! Failing for the first time in my life, deeply shook my self-esteem to the point of doubting my real abilities considering what had thus far been my journey in Bissau: a student who exceeded in assessments and behavior. In this context, I had very low self-esteem and was full of doubts about myself and what I could do.

The support given by his sister was what made him find the strength to overcome this difficult time. She gave him strength when he needed most and made him stay true to his goal.

- *You only failed at one subject: Portuguese. You were almost there! You failed with an 8 out of 20! You know what you're going to do?*
- *I don't know, Sis.*
- *You are not going to give up. You're going to finish your last year! You can do it and you can always count on me to help you with anything.*

That experience shaped him, and that is how he got accepted to ISCSP in 2013, in the International Relations course. That year, he heard several colleagues and teachers share episodes discussed by individuals in the African Students Nucleus, at that time an inactive group — discussing distinctive narratives about "African Week" with its cycle of conferences, African gastronomy shows, exhibitions, and fashion shows. In 2014, he shared with colleagues and teachers his desire to reactivate the Nucleus, to create a structure that, on the one hand, could serve as support for the newly arrived colleagues and, on the other hand, enable the sharing of positivity and diversity of the African continent. That is how, with the collaboration of colleagues and teachers, he assumed the role of president of the ASN between 2014 and 2016.

8.2.1 The Process

Returning to the moment of combined feelings of goodwill and duty — to the "informal conversation in the corridor of a university between a professor and a Guinean student" – we decided that the next step would be to contact the then president of the Nucleus of African Students, Airton César Monteiro (Cape Verdean, 3rd year student of International Relations). So, we did. Hence, the first meeting with African students (the first of several) was an apex. The goal was to realize what could be done to facilitate the integration and the academic success for these students. In a Human Centered Design[5] approach — using brainstorming, word association games, and autobiographies, the following difficulties were identified:

1) Communication in Portuguese
2) Communication in English
3) Study methods and strategies
4) Computing integration
5) Financial issues

These are some of the testimonies collected at the time:

My first day of lectures was tiring because I didn't know where ISCSP was. For me, all the universities in Lisbon were located in Cidade Universitaria (university city), and I had to walk all over Lisbon looking for ISCSP. As soon as I got to the class, I knew I had to make friends and had to sit in the front row to be able to understand what the teacher was saying because he spoke Portuguese so quickly, which made my integration difficult. It took me a lot of effort and dedication. (excerpt from the narrative by Mamadú Saliu, Guinean, student of public administration)

[...] The first difficulty I had was with Portuguese. [...] It is good to point out that I overcame all the difficulties I had thanks to the complementary relationship that existed between Guinean students [...]. (excerpt from the narrative by Bubacar Bari, Guinean, student in public administration)

I joined ISCSP in 2016, studying International Relations. I had a hard time integrating. At first, I found no support, neither from the board nor from my colleagues. I worked very hard to overcome this difficulty. Later I came to have some support from my colleagues from Guinea and Cape Verde. What I went through at university was an experience that I did not want others to go

---

[5]The solution to the problem is designed with the target of the problem in mind. It is also called design thinking, a process characterized by the following moments: empathizing, defining, thinking about the solution, prototyping, and testing.

through, so I am an advocate for more support in the domain of Portuguese and English. (excerpt from the narrative of Eufragio Sami, Guinean, student of international relations)

I'm Anaximandro Monteiro, I'm 23 years old, I'm from Guinea Bissau, I'm a Guinean national, I've lived in Portugal for almost 6 years, with a temporary residency and a study visa [. . .] I speak here in the first person and from my own experience, when I arrived in Portugal I felt many difficulties of a social and technical nature, in fact, I still do today - from integration in the Portuguese community to university results, but little by little I am overcoming the fruit of my persistence. ISCSP welcomed me as well as my colleagues from different PALOP countries, but we have a clear sense of our difficulties (the result of poor preparation in our countries of origin and poor quality of education). Therefore, mine, our concern, is to give us regular attention in order to be able to live up to expectations and overcome barriers, because I believe we are capable. If some succeed, others can also. (excerpt from the narrative by Anaximandro, Guinean, management and human resources student)

At the time, given the location and our available resources, we thought, naturally and linearly, to create support spaces for matters in which students had the most difficulties (Portuguese, English, IT, and study methods). However, we soon faced a question: what about "not feeling integrated neither inside nor outside of the Institution?" If we insisted on that path, probably we would be contributing to accentuate their feeling of not feeling integrated in an academic life. It was in this logic that we started the effort to look for other alternatives, other less conventional solutions. We had to think "outside the box," use lateral thinking.[6] How to involve African students and others? And involve them in what exactly?

We started by trying to find out what others, in a similar situation, would do. We came into contact with several organizations — non-governmental organizations, student organizations, and university projects. The important thing was to identify solutions to identical problems.

Three of the contacts with the greatest impact for the prototype creation phase and who later became partners of Starting Point were the Debate Society of the University of Lisbon (SDUL), AIESEC (global student

---

[6]The term "lateral thinking," created by Edward Bono (in 1967), consists of looking for different ideas, ideas "outside the box." It is the process of solving problems through a creative and indirect approach.

organization dedicated to leadership),[7] and Magna Tuna Apocaliscspiana[8] — 1) because they were already involved and had contacts with other faculties and institutes at the University of Lisbon, they gave us the possibility to "gain scale," reach more students; 2) for its time of existence, the sharing of methodologies, experiences, and purposes; 3) in the case of Tuna, the possibility of integrating students also through music (universal language, capable of uniting different people, from different places, and with different experiences).

Another equally relevant contact was with the "Academia de Líderes Ubuntu" (Ubuntu Leadership Academy), a non-formal education project aimed at training young people with high leadership potential who have come from challenging backgrounds or who have had sincere interest in working there.[9] Even if the target was not the same, the Ubuntu methodology (similar to the AIESEC methodology) provoked a deep reflection around the leaders. Why not think about bringing this perspective to the project we were creating?

In this journey (which lasted approximately nine months), we had the opportunity to read a series of books and articles about today's world — global and VUCA, the centennial generation and the need for soft skills as a necessary instrument for success in almost every aspect of our life. We also had access to tools used in business creation, such as the Business Model Canvas (which we later adopted as one of the Starting Point mindsets).[10]

Basically, what we did during this incubation time was "to think about people who are and think differently" — what Markova and McArthur (2011) call a *collaborative intelligence approach*. What we ended up creating was the Starting Point — IN Educational Experiences (INclusive, IN[non]formal and INspiring), and is thus the result of a process of synergy and interests. Included in this process were our experiences which, what we believe, created a *3rd alternative*, in the words of Covey (2013), a *new box* in Brabandere and Iny (2013), a *cosmopolitan model* for the development of human potential for a global society of tolerance and peace.

---

[7]For more information about this organization, please consult https://aiesec.org/

[8]Magna Tuna Apocaliscspiana is a group of university students, dressed in a traditional academic uniform, who play instruments and sing various traditional songs. The tradition originated in Spain and Portugal in the 13th century as a mean to students to earn money or food. Nowadays, students do not belong to a "Tuna" for this purpose but rather seeking to keep the traditional alive, as well as for fun and to meet new people from other universities.

[9]For more information about this project, please consult https://academialideresubuntu.org/en/

[10]The other, as already mentioned, is the Human Centered Design or Design Thinking.

Before moving on to the presentation and exploration of the model, we consider it important to emphasize that this option for sharing the process itself is due to the fact that we believe that, similarly to what we observed in relation to people — carriers of unique narratives — the projects, are also the result of singular processes and they too contain unique narratives. Thus, understanding the model specifically involves understanding its creation process.

8.2.2 The SP-EEI Model[11]

Like any other model, this must be understood as an image that serves as a reference, as a standard, for similar situations. The originality of this educational model is based upon four factors.

1) Hybridism — results from the combination of two other models (Human Centered Design and Business Model Canvas).
2) Collaborative, synergistic character — solutions that are always thought of together.
3) Flexibility — the option to organize around four social laboratories was the solution found for the problems initially identified. Different problems, quite possibly, would pass for solutions other than these (and the model has this possibility).
4) It is "a means" and not "an end" in itself — the soft skills developed here are a means, a path for the development of more humane people (more knowledgeable and respectful of others) who are better prepared for the labor market in a global and volatile world.

As mentioned previously, the main concern of this project has always abided to the development of human potential, both in terms of skills valued in the job market or in terms of training people to be more human and, as a result, has led to vehicles of peace and tolerance. So, it is understood that our "value position"[11] is precisely *tailor-made* to educational experiences in a cosmopolitan context, addressing triggers to a successful integration and to emotional competence.

It is within this conceptual framework that the *ideation* of the four social laboratories can be understood. They are EmpathyLab, StorytellingLab, CityLab, and VUCALab.

---

[11] SP-EEI is an acronym for Starting Point — Educative Experiences IN.
[11] See Figure 8.2.

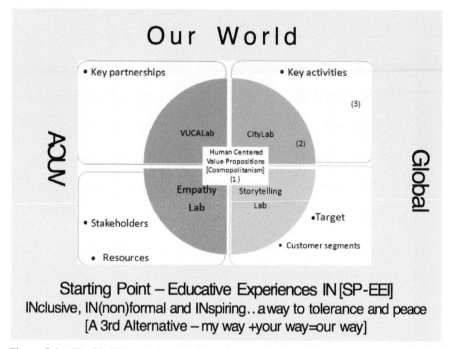

Figure 8.1 The SP-EEI model results from the combination of two apparently paradoxical models — the Human Centered Design, with a focus on the human person and the Business Model Canvas, with a focus on sustainable business. The underlying idea is to make a humanist project sustainable whose mission is to develop human potential. In this case, the solution was to create four laboratories from which all educational activities are designed and organized.

EmpathyLab: The beginning of any relation involves empathy. With these activities, the students have had the opportunity to put themselves in someone else's shoes in order to understand these individuals' attitudes and feelings. Examples of activities include the Human Library and solidarity actions.

Storytelling Lab: Not all stories start with "Once upon a time...." In this lab, students have the opportunity to discover new practices of communicating creatively and in an effective way. Examples of activities include "short storytelling courses," "Book Store Talks" (consisting of personal stories in a bookshop, with inspiring projects led by young people) and public speaking courses.

CityLab: "Lisbon is always a good idea." The goal is to activate the senses — learning to see, to listen, to taste, to smell, to feel, etc. in the city — and with (other-different-and-similar.) Examples of activities involve treasure hunts,

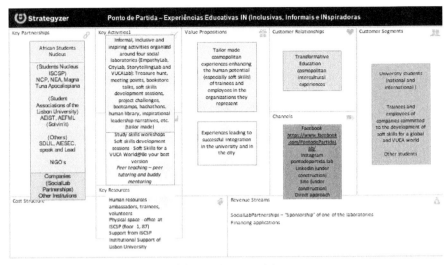

Figure 8.2   Business model canvas.

meeting points (with different themes — music, gastronomy, languages, traditional games, etc.), and "conversations around the table" (a guest, usually a young artist, shares narratives, arts, and some gastronomic flavors).

VUCALab: It provides the development of soft skills in a VUCA and global world so that the student can be "their best version," in multi-cultural spaces. Examples of activities are "Skills for a VUCA World," lectures by young leaders (aiming to the development of skills, with an approximate duration of 1 hour 30 minutes each) "bootcamps," "hackathons" (oriented to reach sustainable development goals), peer teaching, and buddy mentoring.

In regard to the elements designed from the Business Model Canvas, which is also essential to the construction of the project, we can see them in Figure 8.2.

It goes without saying that, despite the short existence of the project, it has been a very enriching experience. Personal social interaction, also called face-to-face interaction, with people who are different from us — (they are not better or worse, they are simply different) who have revealed themselves to be the engine in moments of empathy, compassion, self-knowledge, new friendships, and creativity. Here, we share some sentences from participants in the SP-EEI experiences.

8.2.3 About "What I Enjoyed Most" (Students Quotes)

"You are all so lovely and inspiring. I really loved how you talked about things you do with passion."

"The city tour and all challenges, will help us interact with you guys and others. It was a great experience. I also liked Man talks, which encourages men to express their feelings."

"City game was an opportunity to sit and talk with involved students."

"That we walked together as a group and could share our experience, knowledge and etc. with others."

"Walking in Lisbon, and the talks we had."

"What I liked the most was your energy. Your joy. Activities you prepared for us. Experience you gave us was very important. I'm grateful."

"I liked this meeting because it helped me with getting out of my comfort zone. I'm afraid of meeting and speaking to new people. You are doing a great job. You are really great people and I would like to thank you for that meeting."

"That you are focused on these soft skills that are often forgotten during education. Also, I'm impressed because of your great preparation towards the workshop."

"People, great atmosphere, the way how you were prepared to meet us, to take care of us. It was a really excellent experience. Thank you."

"What I liked most about the activity was having the notion that regardless of our personal difficulties and obstacles, the groups were able to come together, overcoming, for example, language barriers that seemed to be the most prominent in the activity. A good connection between the elements was established in relatively little time, which allowed us to remove from this activity good moments in which, even without having a direct notion of such, we developed several soft skills!"

"Last November 6th, I participated in the activity organized by "Starting Point" I was the leader of Group A - The Smurfs. I really enjoyed taking part in this activity, because I managed in one day to create very strong ties with Dutch students, which seemed like something unimaginable because we didn't know each other before this day. This feeling was the result of the well-organized activities by the co-ordinators. In my opinion, the IN educational experiences (inclusive, informal and inspiring) have been fulfilled. Nowadays, the development of soft skills is very important for any student, including for Political Science students. I am very pleased to have a project as "Starting Point" at ISCSP, which allows the development of this type of capabilities."

8.2.4 About "Ideas that I Take with Me" (Students Quotes)

"I learned that informal ways are essential in improving people, and helps one become useful and successful."

"Learning how to make integration more creative. How to talk to people, how to get closer to them, truly understand them."

"To spread joy to everyone."

"That I should always remember, that we are all equal and everything starts with love."

"The idea to organize Men talks. I'm the organizer of human library, but it's quite different idea."

"That it's good to try to be in a group not alone and be more open for other people and culture."

"Informality of meetings that provide more integration."

"That everyone could learn a lot more if they listened to people from different places and cultures."

"I learned from you about awareness, who I'm and what I can do, that everything depends on me."

"Idea of city walks. I think it's a great way of exploring the city."

"Doing things of importance like social activities. The non-formal way is a good idea and brings a lot of new things, I did 2 pages of notes!"

"Doing everything with Kindness, love."

Having presented the context, the process and the model, it is now time to ask ourselves about the responsibility of universities in creating agents that promote a culture of tolerance and peace. How can models like the SP-EEI become integrated and not just exist at a university? How can they be an integral part of a more socially responsible educational culture?

Having presented the context, the process, the model, and "feedback from experiences," it is now time to situate the model's assumptions within the scope of Education for the 21st century.

8.3 Assumptions of the Model and Pillars of Education — An Analogy

The Report for UNESCO from the International Commission on Education for the 21st Century, chaired by Jacques Delors (1992–1996), is an important contribution to the reflection on the role of Educational Institutions, including the University, in terms of building a culture of peace and tolerance. It is a document of hope, which gives new values to the ethical and cultural dimension of education throughout life.

There are four pillars identified as essential for a successful learning, mainly: learn to know (acquiring instruments of understanding); learn to do (to be able to act on the environment); learn to live together, to live with others (co-operation in all human activities); learn to be (main concept that integrates all of the above).

In terms of facing a global, diverse, complex, uncertain and interdependent world, education is not just used to supply qualified people to the world of economics: it is not intended for human beings as an economic agent but as the ultimate end of development. Developing the talents and skills of each person corresponds, at the same time, to the fundamentally humanistic mission of education, to the demand for equity that should guide any educational policy, and to the real needs of endogenous development, respectful of the human and natural environment, and diversity of traditions and cultures (Delors, 1997: 85).[12]

This is the spirit of the four assumptions integrated by Starting Point — Educational Experiences IN. Let us take a look.

Assumption 1: *We believe that Starting Point's activities, as true cosmopolitan experiences, are places for personal transformation because they mirror the "other," the individual that is different from oneself and those who are closest.*

First of all, it is important to define cosmopolitanism. For Appiah (2008: 13), it is defined as taking the value of not only human life but human lives as a serious subject of great importance, which means taking an interest in the practices and beliefs that give them meaning. People are different, the cosmopolitan is aware of this, and there is much to learn from our differences. Since there are so many human possibilities that are worth exploring, we do not expect or wish that each person or each society will become a unique lifestyle.[13]

Beck stated, regarding the *cosmopolitan moment* we live in, that: "We all live in a direct neighborhood, therefore in a world with others that cannot be excluded, whether we want it to be so or not" (2016: 112). Beck

---

[12] Author's translation.
[13] Author's translation.

further states that more than an issue of *Appiahna* ethics, "cosmopolitanism transforms the inclusion of the other into reality and/or into a maxim" (2016: 113).

In a normative sense ("maxim"), cosmopolitanism means the recognition of cultural differences, both internally and externally. Differences are neither hierarchically classified nor eliminated but accepted as such or even considered positive. However, there is a part of the world in the early 21st century that is far from a situation in which these conditions are accepted. But is there anything that unites people with different skin colors, religions, nationalities, situations, past, and future beyond recognition? The theory of global risk society offers the following answer: traumatic experiences of the community are created forcibly by global risks that threaten everyone's life (Beck 2016: 113).[14]

Thus, in Starting Point — Educational Experiences, "true cosmopolitan activities" are educational and ethically designed for students to learn the meaning of cultural differences and to look at these as something positive, as the scenario of excellence for learning is not just "being" but rather being present. How? This occurs through common projects and objectives.

This assumption is directly related to what is one of the biggest challenges in education — learning to live together, the Third Pillar of Knowledge. For this "relationship-learning" to fulfill its function, it is essential that contact between students is made in an egalitarian context.

If there are common goals and projects, prejudices and latent hostility can disappear and give way to more serene co-operation and even friendship. [...] participation in common projects seems to be an effective method to avoid or resolve latent conflicts (Delors, 1997: 97)[15].

What we observed in the activities carried out was that, as they were united by common goals, national and international students who participated in the activities proved to be highly capable of exploring their similarities, learning from one another. "Everyone could learn a lot more if they listened to people from different places and cultures" (from About "Ideas that I take with me").

It is important to remember that knowledge of the other's culture, or even the recognition of a global identity that brings this sense of belonging, is not in itself sufficient for the student to question his local culture.

---

[14] Author's translation.
[15] Author's translation.

In fact, "when working together on motivating and unusual projects [...] a new form of identification is born, which makes it possible to go beyond individual routines, which value what is common and not differences"[16] (Delors 1997: 98).

In this sense, cosmopolitan places enable the progressive discovery of the other, an effective way of learning without judging, without questioning. It is not easy learning — but who said growing up is easy?

Assumption 2. *We believe that these places, designated here as cosmopolitan places, facilitate essential learning, designated as soft skills that are crucial for personal, social professional and environmental success of millennials, centennials, and other generations.* We had the opportunity to hear from some participants, "(...) The non-formal way is a good idea. And loads of new things, I did 2 pages of notes!" In fact, the presence of the others is an enhancer of human personal development, as it acts as a trigger for the development of skills such as emotional intelligence, empathy, intercultural communication, problem solving, creativity, and critical thinking. It is a privileged way of *learning to be.*

The experience was positive in the sense that I was able to have contact with another culture and I was forced to leave my comfort zone and speak English with people, which was something that I always had difficulty in. Eventually, my English was flowing and I managed to create a pleasant dialogue with the Dutch colleagues (Card4B intern).[17]

We have been aware of the positive effects of non-formal education on the development of human potential for a long time. The added value here is the validation of the importance of non-formal education in the training of more competent people in a university context.

Assumption 3. *We believe that the University, as an institution of higher education, is responsible for producing high-level professionals ready to tackle market challenges and a global world. Moreover, for producing researchers and cultivating human and environmental knowledge. In addition, it also has the responsibility to incorporate and support cosmopolitan programs for developing soft skills in order to guarantee the development of human potential in this Baumanian liquid modernity, which characterizes today's world.*

---

[16] Author's translation.

[17] Company that collaborates with the project as Social Lab Partnership.

Higher education is, in any society, one of the engines of economic development. Traditionally, teaching and research has been the two missions of universities (committed to *learning to knowing*). However, in view of the changes that we have been witnessing, another mission is being considered to reflect the contributions of universities to society. It is generally called the "third mission." This contribution, in general, encompasses three areas: the transfer of knowledge; lifelong learning (that aims to respond to rapid technological developments); and the socio-economic impact on the economic development of the region in which it operates.

In view of the growing and desired internationalization of Universities and their most recent mission — the relationship with society, we believe that the integration of soft skills learning into their curriculum is essential to better adapt them.

Assumption 4. *We believe that a humanist education about, and with, humanity is the path to "constructing the bases" of a culture of tolerance and peace. During this process, it was interesting to understand the role of inspiring narratives and its influence in forming positive beliefs and values* (related with the four pillars).

Regarding the last assumption, now is the time to ask ourselves about the responsibility of universities in creating agents of a culture of tolerance and peace. How can models like the SP-EEIC be an integral part of a more socially responsible educational culture? How can universities contribute to a culture of peace and tolerance?

8.4 "The Responsibility of Universities to Nurture a Culture of Tolerance and Peace"

In order to answer the questions posed previously, we have created an image that reflects our experience and our humble collaboration in this global reflection around paths toward a culture of tolerance and peace. As you can see, this is a multilevel educational approach, as it considers the complexity, multiplicity, and interconnectivity between the dimensions of learning and behavior between various actors at different scales — from the individual to international organizations; from the individual, to planetary consciousness; and from the person | singular group, to humanity.

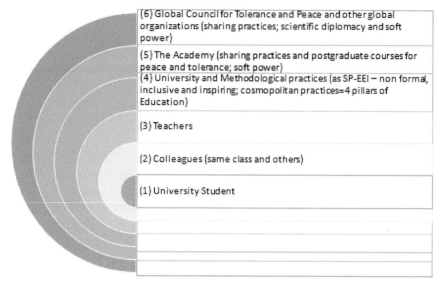

Figure 8.3 The responsibility of the university, the responsibility of all of us.

8.4.1 How Can Models Like the SP-EEIC Be an Integral Part of a More Socially Responsible Educational Culture? (Level 4 of Figure 8.3)

In our perspective, the essence of projects of this nature is the involvement of all students, (national and international). How can we do this? We can achieve this through empathy. In this sense, the Human Centered Design model is shown to be effective — students are involved in the whole process, which gives them both the feeling that they are heard and that they belong to something higher.

The connection with student associations and organizations (internal and external to the university) is also essential: 1) for their experience in non-formal education, 2) for their joy and motivation, and also 3) because these actors are our target (most are still students).

It is also essential to have good support in digital marketing to ensure an effective presence on social networks (Instagram, Facebook, website, etc.). Without this tool, a project of this nature will hardly be able to fulfill its purpose.[18]

It is important to reinforce the idea that in addition to these issues related to structure and organization, if the activities are not truly inclusive and if

---

[18]This is one of the main tensions that we are currently facing.

the practices are not imbued with a cosmopolitan ethics (mentioned in the previous section), we will not be contributing to a socially more responsible society. It is not enough to gather different people in a space and make them play, talk, and have fun. There needs to be a clear purpose with which everyone identifies.

SP is a project that was born in a university; however, it is clear that, for us, the role of these institutions is far from being limited to being the locus of projects like this. When this happens — and it happens more often than would be desirable — the university is not, in fact, contributing to a culture of tolerance and peace. Universities often confuse these methodologies with mere informal activities to welcome international students. As such, they do not benefit from the opportunity to be cosmopolitan spaces where students, involved in common projects and goals, learn to live together, learn to be, and, later, to make society a more humane place.[19]

8.4.2 How Can Universities Contribute to a Culture of Peace and Tolerance? (Level 5 and 6 of Figure 8.3)

Still internally (level 5), and in line with its teaching mission (the transmission of knowledge), several universities offer postgraduate studies in Education for Peace, which is essential for the theoretical and methodological reflection of these themes as well as for the transfer of knowledge to civil society.

However, the university's action on the road to a culture of peace and tolerance is not just internal (nor could it be).

Here, we draw attention to the role of co-operation between scientists, "a powerful instrument for the internationalization of research, technology, concepts, attitudes and activities" (Delors, 1997: 145). Still, in this regard, Varela, Costa, and Godinho (2017) state that scientific knowledge has been used as an instrument of soft power, of inter- and intra-peer negotiation.

Despite this lack of academic attention, co-operation and scientific policies between countries have been of increasing value as a factor in solving global problems. An example of important fields of international scientific co-operation is the fight against climate change, prevention and intervention in the health area [...]. Also, for private agents, scientific and technological knowledge, in addition to functioning as a factor of competitiveness, have been used in co-operation between companies as an asset to influence negotiations and reach new markets. In this sense, we can say that knowledge

---

[19] Approached in Section 8.2.

has been used as a soft power instrument. Soft power is originally defined in international relations as the ability of individuals or a given country to influence others through its virtues (Nye, 2004), which stem, for example, from a country's cultural, musical, sporting, or linguistic influence.

In this context, we can apply these concepts to the theme of science, technology, and innovation (CTI), in which scientific diplomacy is co-ordinated at a national level by the ministries of science and technology of the respective agencies (in the case of Portugal, the Foundation for Science and Technology — FCT) and soft power is managed by institutions linked to research and knowledge, as is the case with Universities and, more recently, other multilateral organizations as we will mention below. Thus, we can say that in the case of scientific diplomacy, academic knowledge, transferred in pedagogical and scientific research terms, functions as an important tool for political soft power (Varela, Costa, & Godinho, 2017: 60).[20]

It is in this sense that we affirm that initiatives such as the one in this book "Paths to a Culture of Tolerance and Peace," under the responsibility of the Global Council for Tolerance and Peace, are fundamental for universities to fulfill their three missions and, in this way, be useful to society.

In conclusion, it is likely that, during the reading of this article, perhaps you have come across and established some parallelism between the assumptions and practices of the model presented with other models and other practices, especially those that are related with education for peace, tolerance, or even global citizenship. After all, the concern with contributing to the construction of a culture of peace and tolerance is not new. We have even gone so far as to affirm that the ideation of the model presented, in the first phase, involved a careful look at "other boxes" (Brabandere & Iny 2013), that is the models, concepts, and methodologies that already exist and that serve similar purposes. So, our creative process went through the creation of a "new box" — the university context, but now from the "outside" (non-formal education) and with "bridges" — a common project — that connect this place to other places that can be nearby or even far away, but all in a diverse, global world.

It is thus a method of learning to be a global peacemaker. Still, in this regard, we can look at what Delors states:

"This organization [UNESCO] will be serving peace and understanding among men, by valuing education as a spirit of harmony, of the emergence of a desire to live together as militants of our global village that must be thought

---

[20]Author's translation

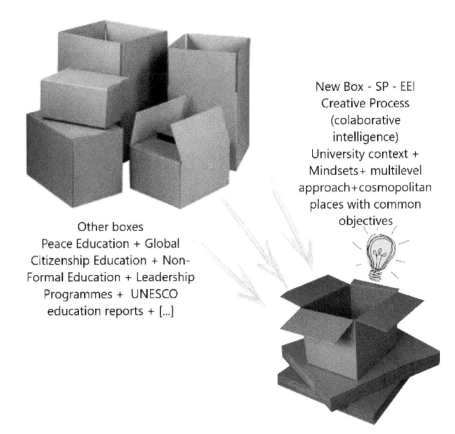

New Box - SP - EEI
Creative Process
(colaborative
intelligence)
University context +
Mindsets+ multilevel
approach+cosmopolitan
places with common
objectives

Other boxes
Peace Education + Global
Citizenship Education + Non-
Formal Education + Leadership
Programmes + UNESCO
education reports + [...]

Figure 8.4   We consider this metaphor of a "new box" interesting because it communicates the idea that more than creating, what we do is recreate — adapt what already exists to the context and purposes in question. In this process, it is essential to take into account existing resources (present) and the sustainability of the project (future). The line that combines these elements — the idea, the purpose (= assumptions), the resources, and the sustainability — is the same mindset (in this case, as mentioned in the previous section, it results from the intersection between Human Centered Design and Business Model Canvas). Collaborative intelligence, the way we communicate with not only others but all agents, at different levels.

and organized, for the benefit of future generations. In this way, you will be contributing to a culture of peace." (Delors, 1996: 31)[21]

Universities, on the one hand, with the creation of a social space for their students to learn to know, to do, to live together and to be, and, on the

---

[21] Author's translation.

other, with the management of their political soft power within International Relations, will certainly be contributing to a culture of peace and tolerance.

References

Appiah, K. (2008). *Cosmopolitismo. Ética num mundo de estranhos.* Publicações Europa-América.

Beck, U. (2016). *Sociedade de Risco Mundial.* Lisboa: Edições 70.

Brabandere, L. d. & Iny, A., 2009. Thinking in New Boxes. How to bring fundamental change to your business.. *The Boston Consulting Group,* pp. 1–3

Covey, S., 2013. *A Terceira Alternativa.* 2a edição ed. s.l.:Gestão Plus.

Delors, J., 1997. *Educação. Um tesouro a descobrir. Relatório para a UNESCO da Comissão Internacional para o Século XXI.* São Paulo: Cortez Editora.

Kant, I. (2018). *A Paz Perpétua e Outros Opúsculos.* Lisboa: Edições 70.

Kant, I. (2019). *Fundamentação da Metafísica dos Costumes.* Lisboa: Edições 70.

Markova, D. & McArthur, A., 2011. *Collaborative Intelligence. Thinking with people who think differently.* 1st ed. New York: Spiegel & Grau.

Nussbaum, M. (2003). Capabilities as Fundamental Entitlements: Sen and Social Justice. *Feminist Economics, 9 (2/3) ,* 33–59.

Nussbaum, M. (2012). *Women and Human Development. The Capabilities Approach.* Cambridge Univesity Press.

Ozdemir, I. (24 de Abril de 2016). *Peace and Tolerance through Education - A New Role for Universities.* Obtido em 17 de Dezembro de 2019, de Global Campaign for Peace Education: http://www.peace-ed-campaign.org/peace-tolerance-education/

Popkewitz, T. (2009). *El Cosmopolitismo Y La Era De La Reforma Escolar .* Morata.

Reardon, B. (2001). *Education for a culture of peace in gender perspective.* França: Unesco Publishing.

Reardon, B., & Snauwaert, D. (2011). Reflective pedagogy, cosmopolitanism and critical peace education for political efficacy: A discussion of Betty A. Reardon's assessmemnt of the field. *Factis PAX, Vol.5, N 1 ,* 1–14.

Roth, K. (2015). The role of examples, current designs and ideas for a cosmopolitan design of education. *Policy Future in Education, vol.13 (6),* 763–774.

Sen, A. (2009). *The Idea of Justice.* Harvard University Press: Cambridge, MA.

Sen, A. (1985). *Commodities and Capabilities.* Amsterdam: North-Holland.

Serres, M. (2019). *Tempos de Crise.* Lisboa: Guerra e Paz.

Singer, P. (2019). *Ética no Mundo Real.* Lisboa: Edições 70.

Varela, C., Costa, C. & Godinho, M., 2017. Diplomacia científica: do conhecimento académico ao soft power político. *JANUS,* pp. 60–61.

9

The Peace Education Reconstructive-Empowering Approach: From Recognition to Cultures of Peace

Dr. Sofia Herrero Rico

University of Jaume I

Abstract

This article aims to motivate the reflection, awareness, and empowerment for positive action to all those interested in education and in the conviction that it can be addressed to reduce the existing intolerance, exclusion, and violence and contribute to cultures of peace.

This will be possible by deconstructing the direct, structural, and cultural violence, which is reproduced in the educational system, as a social subsystem, and reconstructing different ways of doing peace education, specifically through the Reconstructive-Empowering (REM) approach. In this task, recognition of the others will be a key element to produce positive social transformation and to foster cultures of peace.

Keywords: Education, empowerment, competence, recognition, cultures of peace.

9.1 Introduction

In this article, I take the opportunity to reflect on the Peace Education (hereafter PE) Reconstructive-Empowering (hereafter REM) approach (Herrero 2019; 2018; 2013; 2012; 2009) that I propose from my research in the UNESCO Chair of Philosophy for Peace (CUFP) and at the Interuniversity

Institute for Social Development and Peace (IUDESP) of the Universitat Jaume I of Castellón (UJI), Spain.

The PE REM approach takes as starting point the hypothesis that humans have capabilities and competences to make peace. We are not determined to be genetically violent, but it depends more on our environment and culture. Therefore, violence and destruction are avoidable. Since we have the same capabilities to behave peacefully or violently, our response depends in the long run on the education we receive. Therefore, humans are responsible for creating one kind of behavior and not another. Thus, we highlight the importance of PE. With this regard, we state that we can make peace among us as well as with nature (Martinez, 2001, 2005; Adams, 1992). We, human beings, have the "power" to make peace and to transform our conflicts by peaceful means. PE, understood from the REM approach, is the reconstruction of these abilities and competences. This education approach is understood under the dialogic–participative paradigm (Martin, 2010). Every human being is considered a valid interlocutor and capable of making peace. Through recognition and peaceful interpellation, we can be able to create understanding and coexistence and, therefore, cultures of peace.

For the development of this article, I will divide the text into different sections and subsections to give greater coherence and clarity. Therefore, the sections in which I will distribute the content are the following: PE from the REM approach, overcoming intolerance and violence through recognition, and the proposal of recognition as an engine for the creation of cultures of peace.

9.2 Peace Education from the Reconstructive-Empowering Approach

The PE REM approach takes as a conceptual basis the *philosophy of making peace* of Martínez Guzmán (2001, 2005, 2009) from which PE is interpreted as the reconstruction of our human competences, in the sense of capacities or powers, to make peace. On the one hand, I call it *reconstructive* because it focuses on the reconstruction of our skills and abilities to make peace, which we have as a human characteristic. On the other hand, I refer to it as *empowering* because it requires awareness, motivation, and recovery of those powers we have for the peaceful transformation of the daily life conflicts, which are inherent to human relationships (Martínez, 2001; 2005). Thus, the REM approach demands the reconstruction of our human competences to make peace and our empowerment to bring these peaceful competences to action, from our daily experiences. The aim is to educate in the difference and in the conflict that it entails, and by not facing it with violence but rather by peaceful means, turning them into learning opportunities (Paris, 2009).

In this sense, the elements of the PE REM approach will be distributed in response to the three components that we have to consider when implementing any educational practice and/or action, including PE (Cabezudo and Haavelsrud, 2001: 279). Therefore, this PE proposal includes the following vectors.

1) What should be taught? This will respond to the selection of the contents, which we briefly name as follows:

   - A plural, diverse, holistic, and positive concept of peace (Herrero, 2019, 2013).
   - The responsibility and social competence to make peace from our personal and everyday experiences (Marina and Bernabeu, 2007).
   - The ability to change our perceptions and perspectives in order to put ourselves in the place of the other party, empathize and understand that, on many occasions, different visions, beliefs, judgments, and interpretations are possible depending on the perspective (Strawson, 1995).
   - Communication for peace (Nos, 2007) and non-violent communication (Rosenberg, 2017) as a methodological tool to allow us, through performativity and our ability to change discourses, behaviors, attitudes, and perceptions by putting them in dialogue with the others. Therefore, through mutual interpellation, others can demand us accountability if what we express through what we do, what we say, and even what we keep in silence is not accepted, and, thus, we will be able to reach agreements.
   - The importance of cooperation for the peaceful transformation of conflicts. Cooperation is required and needed if both sides want to overcome the conflict since it is necessary to cooperate with the other party in order to make peace (Rapoport, 1992).
   - The empowerment to bring peace to action. Empowerment is understood as the revaluation of our powers and abilities to make peace as a human characteristic (Bush and Folger, 1996; Lederach, 1984, 1996; López, 2006; Muñoz, 2000).
   - Care ethics (Comins, 2009) interpreted as a human value and not only as a gender characteristic, since it is necessary to learn to

take care of others, as well as of nature to contribute to cultures of peace. It also includes sentimental coeducation (Comins, 2009), meaning that we must educate girls and boys equally, including the expression of feelings and emotions in a natural and positive way.

The recognition of all human beings as valid interlocutors (Honneth, 1997, 2008, 2011). We interpret the concept of recognition as a step beyond tolerance since recognition takes as a starting point the consideration of other human beings as equally valid to oneself, at the expense of their physical, geographical, and cultural differences. To address this concept more broadly, I will dedicate a section on its own, as I consider it as a key concept and the engine of the creation of cultures of peace, which is the ultimate objective of this article.

2) How can this be taught? This would correspond to the form or methodology used in the teaching–learning process. Thus, the methodology that implements the REM approach would be that of deconstruction–reconstruction, that is, "we will unlearn what we have learned badly or stopped learning due to the rigidity, authoritarianism and violence of our educational-social systems" (Herrero, 2012: 53). In this sense, we will deconstruct the direct, structural, and cultural violence classified by Galtung (1985, 1993) that reproduces the educational system as a social subsystem and we will rebuild an education based on peace values and the training of our potential (powers) or competences to make peace from our personal experiences and everyday life.

3) Where can this be taught? That would relate to the contextual conditions that are required to implement the PE REM approach. On this third issue, a dialogic, inclusive, communicative, free, dynamic, and interactive context is required, highlighting the 2.0 contexts taking us into the jargon of cyberspace (Martin, 2010). Of course, contextual conditions must have the characteristics of a culture of peace (Adams, 1992). Likewise, the context of the PE REM approach should not be limited to the formal context of the educational system, but it should also cover informal contexts such as family or media; and non-formal ones, for instance, other entities with which we interact and educate ourselves, such as sports, leisure, and cultural associations, among others.

9.3 Facing Intolerance and Violence Through Recognition

Education, as a social subsystem, reproduces the three types of violence proposed by Galtung (1985, 1993): direct, structural, and cultural. In general, education is understood under the narrative–passive paradigm (Martin, 2010), as a banking education in words of Freire (1970), with a greater focus on individuality and competition rather than on cooperation and understanding of each other to promote peaceful coexistence. Considering this, we can name the following examples of different types of violence.

1) *Direct violence* (this violence is the consequence of the use of force or violence in any external manifestation). Education reproduces different kinds of direct violence such as labels, contempt, insults, punishments, shouts, blows, bullying, and fights, among others.

2) *Structural violence* (this violence is the consequence of an unequal process of social construction and socialization, which are inherited in our social structures and systems (e.g., political, legal, economic, cultural, etc.), and provokes human stratification). Education, as a social subsystem, also reproduces different kinds of structural violence. For instance, we can see social division of labor (manual and intellectual), educational hierarchy, abuses of power and coercion, vertical relationships, unidirectional and vertical communication, competition, a lack of equal opportunities for all, and exclusion, among others.

3) *Cultural violence* (this is a symbolic violence which is expressed by infinite means — religion, ideology, language, art, science, media, education, etc. — and fulfills the function of legitimizing direct and structural violence, as well as inhibiting or repressing the response of those who suffer it). Examples of cultural violence in education can include xenophobia toward others — prejudices, stereotypes, gender roles, cultural and anthropological ethnocentrism, extremism, and racism, among others.

Considering the three types of violence proposed by Galtung (1985, 1996), I would like to put him in dialogue with Honneth (1997, 2008, 2011), who defines the three most common forms that we tend to use to despise others. Therefore, Honneth (2011) classified three ways of contempt, which can be addressed as well as the three types of violence of Galtung.

1) The contempt of the body or the physicist (direct violence). The non-recognition of the body and of the physical conditions causes, in the

injured people, the alteration of their identity and the loss of trust in themselves.

2) The contempt of the rights of a certain person (structural violence). The non-recognition of the human rights and of the people legal aspect causes, in the injured people, the loss of self-respect because the person is considered excluded from the juridical and moral community.

3) The contempt for the culture or way of life of the others (direct violence). The non-recognition of the different cultures and lifestyles causes, in the injured people, the feeling of exclusion, marginalization, and underestimation and, consequently, the loss of self-esteem.

By combining the three types of contempt I have just mentioned, we could reflect, for example, on situations in which we exclude or marginalize other people for being disabled, fat, dwarf, black, gypsy, immigrant, poor, refugee, prostitute, lesbian, transsexual, punk, transient, indigenous, etc. As we can see, contempt causes in the aggrieved people a lack of trust, lack of respect, and lack of self-esteem by generating the different types of violence, and, therefore, we can conclude that the non-recognition of others can lead to cultures of violence.

With this regard, in order to face intolerance, contempt, discrimination, and violence, Honneth (1997) proposed the struggle for recognition.

9.4 Recognition: Key Concept for the Creation of Cultures of Peace

The PE REM approach foments the recognition (Honneth, 1997, 2008, 2011) of all human beings as valid interlocutors, considering every single being as equally valid with special attention to her/his culture, social class or religion (Martínez, 2001, 2005). We can interpret recognition as one step further from tolerance because recognition focuses on overcoming ethnocentrism, stereotypes, and prejudices by considering all people as valid interlocutors, on a level of equality and despite the fact that you share or not the others' ideas, cultural forms, religious manifestations, or ways of understanding and living life.

According to the Spanish Dictionary *Diccionario del Uso de Español* by María Moliner (2009) "recognition" is the action of recognizing, defined with the following acceptations:

1) To be aware that one person or thing is precisely one determined, known, and identified.

2) To admit that a certain person is what he/she expresses and to recognize him/her with his/her legality and authenticity.
3) To recognize that a certain thing or person exists and has their own value even if it/they dislike me.

The three acceptations of the definition made by Moliner (2009) are very important for the fact of recognizing others and accepting diversity as an enriching element of our society. Following Honneth (1997, 2008, 2011) and, as I have mentioned in the previous chapter, we refer to recognition by taking into account the three main types of disrespects we do to others: disrespects related to our bodies or appearances, disrespects related to our legal and human rights, and the disrespects related to our cultures, religions, or lifestyles. Therefore, recognition is required to overcome intolerance and violence and to contribute to the creation of cultures of peace. Recognition is overall seen in the following three aspects:

1. The recognition of the body, which promotes esteem, care, love, and self-trust.
2. The recognition of the legal and human rights, which promotes identity, integration, solidarity, empathy, and self-respect.
3. The recognition of the different lifestyles, which promotes self-esteem.

Therefore, the PE REM approach focuses on the reconstruction of self-trust, self-respect, and self-esteem through the three forms of recognition proposed by Honneth (1997, 2008, 2011). In sum, the recognition of these three dimensions — that is physical, legal, and cultural — will overcome the culture of violence and will contribute to the creation of cultures of peace.

9.5 Conclusion

The PE REM approach invites us to reflect on what logic and rationality we have established as human beings and on which education is based. Apparently, from politics, mass media, dominant culture, and education, the logic of violence, individuality, competitiveness, contempt for difference, intolerance, and exclusion are mostly shown. Thus, the educational system, as a social subsystem, also inherits this logic by reproducing not only direct violence but also structural and cultural violence, which are more subtle and difficult to make visible. Therefore, it is necessary to modify these logics and deconstruct existing violence if we want to educate for peace and create a more respectful, tolerant, and inclusive society.

Through the reading of this article, we see that we have powers and competences, as well as different alternatives to making peace. We know, then, that "violence is not a biological fatality included in our genes, but that it is learned through processes of socialization and acculturation, in the same way that we can learn nonviolence and peace" (López, 2006: 71). However, it is in our hands to unlearn the culture of violence and war (Bastida, 2004), which is the consequence of not recognizing the otherness and to learn the cultures of peace through recognition and peaceful interpellation. We, as human beings, are responsible for contributing to a more peaceful coexistence from our personal and daily experiences.

In this endeavor, the PE REM approach, which is based on the recognition of others — not only on the physical integrity of people but also on their human and legal rights and of their different cultures and ways of life — aims to be an alternative to contribute to the creation of cultures of tolerance and peace.

References

Adams, D. (1992). El Manifiesto de Sevilla sobre la Violencia. En Hicks, D. (ed.), *Educación para la Paz. Cuestiones, principios y prácticas en el aula (p. 203–295)*. Madrid: MEC/Morata.

Bastida, A. (2004). *Desaprender la guerra. Una visión crítica de la Educación para la Paz*. Barcelona: Icaria

Bush, R. & Folger, J. (1996). *La promesa de la mediación. Cómo afrontar el conflicto mediante la revalorización y el reconocimiento*. Barcelona: Granica.

Cabezudo, A. & Haavelsrud, M. (2001). Rethinking Peace Education. En Webel, Ch. y Galtung, J. (eds.), *Handbook of Peace and Conflict Studies* (p. 279–296). New York: Routledge.

Comins, I. (2009). *Filosofía del Cuidar*. Barcelona: Icaria.

Freire, P. (1970). *Pedagogía del oprimido*. Montevideo: Tierra Nueva.

Galtung, J. (1985). Acerca de la Educación para la Paz. En Galtung, J. (ed.), *Sobre la Paz* (p. 27–72). Barcelona: Fontamara.

Galtung, J. (1993). Paz. En Rubio, A. (ed.), *Presupuestos teóricos y éticos sobre la Paz* (p. 47–52). Granada: Universidad de Granada.

Herrero, S. (2019): ≪La Educación para la Paz desde el poder y la competencia: el enfoque REM≫ en Cabello-Tijerina, P., G. diaz Pérez y R. vazquez-gutierrez (eds.) (2019): *Investigación para la Paz: Teorías, Prácticas y Nuevos Enfoques*, Valencia, Tirant Lo Blanc

Herrero, S. (2018). ≪Education for nonkilling creativity: a reconstructive-empowering approach≫, *Journal of Peace Education,* 15 (3), p. 309–324, ISSN: 1740–0201, Online ISSN: 1740–021X: Routledge

Herrero, S. (2013). *La Educación para la Paz. El enfoque REM (Reconstructivo-Empoderador).* Publicia: Alemania.

Herrero, S. (2012). Educando para la paz a través del reconocimiento de la diversidad. En Nos Aldás, E., Sandoval Forero, E. y Arévalo Salinas, A. (eds.), Migraciones *y Cultura de paz: Educando y comunicando solidaridad (p. 41–56).* Madrid: Dykinson.

Herrero, S. (2009). ≪La Educación para la Paz desde la filosofía para hacer las paces. El modelo Reconstructivo-Empoderador≫. En París Albert, S. y I. Comins Mingol (eds). *Filosofía en acción. Retos para la paz en el siglo XXI,* (p. 33–57), Castellón: Universitat Jaume I.

Honneth, A. (1997). *La lucha por el reconocimiento. Por una gramática moral de los conflictos sociales.* Barcelona: Crítica.

Honneth, A. (2008). *Reification. A New Look at an Old Idea.* New York: Oxford University Press.

Honneth, A. (2011). *La Sociedad del Desprecio.* Madrid: Trotta.

Lederach, J. P. (1984). *Educar para la paz. Objetivo escolar.* Barcelona: Fontamara.

Lederach, J. P. (1996). *Preparing for Peace. Conflict Transformation Across Cultures.* Nueva York: Syracuse University Press.

López, M. (2006). *Política sin Violencia. La Noviolencia como humanización de la política.* Colombia: UNIMINUTO.

Marina, J. A. & R. Bernabeu (2007). *Competencia Social y ciudadana.* Madrid: Alianza.

Martín, M. (2010). Ciencia, Tecnología y Participación Ciudadana≫. En Toro, B. y Tallone, A. (Coord.), *Educación, Valores y Ciudadanía* (p. 41–57). Madrid: Organización Estados Iberoamericanos para la Educación, la Ciencia y la Cultura (OEI).

Martínez, V. (2001). *Filosofía para hacer las paces.* Barcelona: Icaria.

Martínez, V. (2005). *Podemos hacer las Paces. Reflexiones éticas tras el 11-S y el 11-M.* Bilbao: Desclée De Brouwer.

Martínez, V. (2009). *Filosofía para hacer las paces.* Barcelona: Icaria.

Moliner, M. (2009). *Diccionario del uso del español.* Madrid: Gredos.

Muñoz, F. A. (2000). El Re-conocimiento de la Paz en la Historia. En Muñoz Muñoz, F. A. y López Martínez, M. (eds.), *Historia de la Paz: Tiempos, Espacios y Actores.* Granada: Universidad de Granada.

Nos, E. (2007). *Lenguaje Publicitario y Discursos Solidarios*. Barcelona: Icaria.

París Albert, S. (2009). *Filosofía de los Conflictos*. Barcelona: Icaria.

Rapoport, A. (1992). *Peace. An Idea Whose Times Has Come*. Ann Arbor: The University of Michigan Press.

Rosenberg, M. (2017). *Comunicación NoViolenta. Un Lenguaje de Vida*. Barcelona: Acanto

Strawson, P.T. (1995). Libertad y resentimiento y otros ensayos. Barcelona: Paidós

10

The Meaning of Education in a Time of "Ressentiment" and Global Hatred

Prof. Stephen Dobson

Victoria University of Wellington, New Zealand

Abstract

"The most important demand placed upon all education is that Auschwitz [does] not happen again" - these are the words of philosopher and sociologist Theodor Adorno, a German who gave a famous radio talk in 1966. This chapter presents and updates this important discussion with reference to selected recent global events and offers a discussion of *ressentiment*. I argue that in educational practice, it is important to understand what *ressentiment* is and how to teach about it as the learning of the unsociable-social and its counterpart, *non-ressentiment*, and the sociable-social.

A simple question is asked in this chapter: what is the meaning of education? On the one hand, this is a question that, if left too general, lacks connection with the lives in which we live. To counter this, I will reflect upon the meaning of education in the wake of two particular events, by no means unique to our times: the terrible massacre carried out by a lone gunman in Norway on 22 July 2011 and the terrible massacre carried out by a lone gunman in New Zealand on 15 March 2019. We might, on the other hand, turn to the other extreme and answer with reference to the detail of curriculum or carefully selected cross-curricula skills, such as cooperative learning in teams. This narrowing would mean we might lose the ability to move between multi-level explanations drawing upon socio-political, educational, historical, cultural, and psychological explanations.

In answering this question, I am also conscious of how the nature of knowledge seems to be in flux now more than ever before. For some, such as those inspired by Siemens' (2005) seminal paper, knowledge is distributed widely in different networks, some conceptual — carried in our heads — and some external in books or on the Internet, and it raises the important questions of where, how, and what knowledge is to be trusted, taught, acquired and in what settings. He proposed we turn to the concept of *connectivism* to address these concerns and wrote:

> "The pipe is more important than the content within the pipe. Our ability to learn what we need for tomorrow is more important than what we know today. A real challenge for any learning theory is to actuate known knowledge at the point of application. When knowledge, however, is needed, but not known, the ability to plug into sources to meet the requirements becomes a vital skill. As knowledge continues to grow and evolve, access to what is needed is more important than what the learner currently possesses."

Siemens' point is simple,[1] education will increasingly be about teaching and learning that is able to connect together different sources and networks of knowledge, residing in particular places and repositories - sometimes in the heads of others, sometimes recorded elsewhere. What we need is actionable knowledge where critical thinking is still vital but of the character required to select and evaluate the pipe and the contents of the pipe. Such knowledge adds to the epistemological view that knowledge is not merely about "know this" and "know how", but it also entails "know where" and "know how it feels". The last mentioned is essential, ensuring that knowledge is embraced as comfortable and meets our expectations and shared norms of acceptability.

For this reason, in this chapter, I remain wedded to the ideas of connectivism and multi-level explanation but draw upon narrative knowledge that connects knowledge with lived experience communicated in the form of narratives. The turn to stories and storytelling is not uncommon in the social sciences and it represents the attempt to understand how knowledge can take many forms that can be valid and trustworthy. It is not the case that only natural science holds the golden key to truth. As Ricoeur (1984: 3) put it:

> "Time becomes human time to the extent that it is organised after the manner of a narrative; narrative, in turn, is meaningful to the extent that it portrays the features of temporal existence."

---

[1] In this section of the chapter, I am drawing upon unpublished ideas co-developed by Edward Schofield and me in the short paper: Keep it simple, coordinated and normal – the rush to "online-ness."

This phenomenological understanding is circular, as life and narrative mirror each other in a creative mimesis. It also places an emphasis on the necessity of the narrative revealing the ordering of the events in what is basically a linear temporal "causal sequence" (Ricoeur, 1984: 41). However, Ricoeur ignores postmodern and hypertext-inspired conceptions of narratives that break with the temporal organization of the plot in a diachronic beginning-middle-end and disrupt the direction of causality (Boje, 2001). In my work with refugees, I found instances of narratives where a single narrative beginning was unclear, or an authorship could not be traced to a single origin. Instead narratives were multi-punctual in origin and multi-accented because of the polyphonic presence of several voices (Dobson, 2004: 131–134). It is worth noting that in such cases, causality was not necessarily absent but multi-accented and/or multi-directional. A multi-directional narrative can be defined as a narrative that proceeds forward as well as backwards in search of an origin. This reversal of the causality means that it is not cause to effect, but an effect or several effects in search of a cause and this becomes the focal point of the narrative moving backwards.

Accordingly, to understand the meaning of education in a time scared by violence, I shall tell three connected stories in this chapter.

1. The story of Norway and New Zealand in old and new; that is, before and after these atrocities.
2. The most important demand placed on all education, where I recall Adorno's rightly famous essay, originally given as a radio talk with an openly multi-level approach to explanation.
3. *Ressentiment,*[2] moral, and values-based education; covering different cultural and transcultural understandings of education in a time of global hatred.

Broadly speaking, the first story considers what happened, the second why, and the last, how might we avoid that it happens again through education. All good narratives contain the how, the why, and the what if, with a clear line of connection drawing them together.[3]

---

[2]In talking of ressentiment in an applied educational sense, I shall draw upon ideas first co-developed by Dobson and Halland (1995).

[3]I am riffing off Aristotle (1965) in *On the Art of Poetry*, where he proposed that a narrative involved a beginning, middle, and end organized in a causal direction so that events are joined together to reveal a plot.

10.1 Story Number 1: Norway and New Zealand, the Old and the New

Both countries share the history of these terrible atrocities — first some context. Norway is a country on the edge of Europe with 5.2 million in population and it received independence from Denmark in 1814 and from Sweden in 1905. It is a long, stretched out country, a bit like Italy in shape. In the north, we find the traditional home of the Sami indigenous people who are known for reindeer herding. They have been subjected to Norwegian rule for hundreds of years, but today, they have their own Parliament, founded in 1989 with responsibilities for Sami politics, culture, language, and different funds. The Sami are also found in the neighboring countries of Finland, Sweden, and Russia. Norway is known for its social democratic traditions and the Nobel Peace Prize awarded each year in Oslo. Once the poorest country in Scandinavia, it has since the discovery of oil and gas in the North Sea in the 1970s, reversed this position.

New Zealand shares many similarities with Norway despite the smaller geographical size. It is on the periphery of South East Asia with mountains, snow, wind, forestry, love of milk products, fish, oil and gas reserves (of a lesser scale than Norway), and a colonial history. The country is founded on the *Treaty of Waitangi* (1840) and this states the principles of partnership, participation, and protection between representatives of the British Crown and the Māori people, who constitute 15% of the population. The implementation of these principles still remains the source of discussion and controversy. New Zealand only became a sovereign entity with control over its constitutional arrangements and foreign affairs in 1947, and actually waited until 1987 to become a free-standing constitutional monarchy with a parliament possessing unlimited sovereign power.

The old Norway never believed an act of terror could take place in their country. Who would even identify the country as a target? That was until 22 July 2011 on a Friday when the country was undertaking its annual summer vacation. A Norwegian parked a non-descript car outside the tall building in which the Prime Minister had his office and left. As the bomb exploded, he was driving to Utøya Island in the middle of a small lake. Now dressed as a policeman, he made his way to the island, where the annual summer camp for the Labour Youth League was taking place, and executed many people. In total, 77 people died that day. His rationale was to remove the next generation of Norwegian labor politicians who would, in his view, be responsible for immigration into the country. The day after this event in every street and in every household, there was silence. Later, the Norwegian Prime Minister, Jens Stoltenberg, phrased it as, "it was one Norway before and one after 22 July". The new Norway had been born.

The old New Zealand never believed an act of terror could take place in their country. There had been extreme violence in the New Zealand Wars (Ngā Pakanga o Aotearoa) between different Māori groups and the British Crown, but for many, this was an event in the distant 19th Century (O'Malley, 2019). During Friday prayer in the city of Christchurch, on 15 March 2019, a lone gunman proceeded to attack the Masjid Al Noor and the Linwood Islamic Centre mosques a short distance away. Jacinda Ardern, the New Zealand Prime Minister, talked on the day of those wounded and the 50 who died (one would die later because of the injuries):

> "Many of those who will have been directly affected by this shooting may be migrants to New Zealand, they may even be refugees here. They have chosen to make New Zealand their home, and it is their home. They are us. The person who has perpetuated this violence against us is not. They have no place in New Zealand. There is no place in New Zealand for such acts of extreme and unprecedented violence, which it is clear this act was."

New Zealand was in total shock. In Parliament, the following week, the Prime Minister said:

> "He will, when I speak, be nameless. And, to others, I implore you: speak the names of those who were lost, rather than the name of the man who took them. He may have sought notoriety, but we in New Zealand will give him nothing. Not even his name."

For many weeks it was so, his name was not spoken in the public space. In the aftermath, gun laws were changed at a rapid pace beyond the comprehension of many outside of the country. By 10 April, the Government passed the Arms Amendment Act that banned semi-automatic firearms, magazines, and parts. By 22 July, over 2000 guns had been handed in at buyback events. The new New Zealand had been born.

10.2 Story Number 2: The Most Important Demand Placed on all Education

So far, I have told a story about what happened and as a bridge to Story 2 and how to account for the actions of the two perpetrators, I note that in both countries, what to tell children, who should do this, and with what support was quickly raised. Advice to parents was supplied by psychologist experts in child trauma after accidents and societal catastrophes. There was also talk of how to support teachers in their first meetings with children after the event. It was summer vacation in Norway, but not so in New Zealand where schools were open on Monday. This was the challenge in the short term and professionals were available in both countries to support parents, teachers, and children. In the longer term, the attention turned to the questions: how could this happen; was there something amiss or faulty in the education of children and in the education of these two perpetrators in particular?

With this in mind, in this story, I will seek to understand how the two perpetrators came to undertake these terrible acts and what should be the purpose of education in more general terms for all. I will follow in the footsteps of Adorno's and begin with a brief retelling and commentary of a radio talk he gave in 1966 entitled *Education After Auschwitz*. There are obvious comparisons in his topic and those involved in Auschwitz. How could they have been educated to undertake these terrible atrocities, and what kind of education and societal events equipped them? A multi-level approach to explanation is required, referencing socio-political, educational, historical, cultural, and psychological factors. Reducing the cause to only individual factors would negate all the other forces that form and influence the individual.

Adorno opens with the assertion, "the premier demand placed upon all education is that Auschwitz [does] not happen again". He immediately directs attention to the question, so why were many Germans disposed to supporting Hitler? The humiliation of the Germans at the end of World War I and the hyper-inflation of the 1920s meant many were disappointed with politicians. But it is hard to simply change through political means what were in many respects the effects of a global economic depression and having lost the war. This said, these factors had a visible effect: children growing up would have seen the weakness and impotence of their fathers to put food on the family table. The ground was fertile for what psychiatrists would call replacement father figures. Hitler was waiting in the shadows and gradually occupied the political stage in this respect during the 1930s.

Hitler had an educational project. To recover the greatness of the German people, it was necessary to train the youth, and the vehicle for this was the Hitler Youth organization (*Hitlerjugend*), actually founded in 1922. The activities of this group focused on physical and also emotional resilience, and Adorno makes the point that such an education was an important foundation for educating individuals to hardness, a lack of empathy, and a willingness for those who persecuted the Jews and other groups to treat them as objects and not human subjects. It meant they were equipped to kill them when the time came without regard for their human right to live. With this background tapestry, Adorno concludes that while the content of knowledge is important across the different disciplines, it means nothing if school children have no moral understanding of others and how this knowledge might be used to reach ends that are far from humane and noble.

In his view, students in the time of Hitler were led by strong replacement father figures and who had not been educated to think independently and critically. He was concerned that new examples of Auschwitz would take place if education failed in this educational task, where additionally acting and reacting humanely constitutes the moral compass. For Adorno, in the practice of education, we must continually raise awareness of the conditions that supported monstrosities, such as Auschwitz.

So, if this is the "premier demand placed upon all education", how can we understand what went wrong with the upbringing of the perpetrators of the events in Norway and New Zealand? As with Adorno's analysis, the blame cannot be placed solely upon a single factor. Seierstad (2019) has noted that both craved notoriety, as evidenced by the 1500-page cut-and-paste manifesto written by Breivik and the 74-page equivalent by the New Zealand gunman. It is further reflected in their narcissistic personality disorders — Breivik diagnosed by the court psychiatrist and the New Zealand gunman showing the same traits. They both displayed a mixture of rage against Muslim immigrants and self-pity as victims.

Breivik grew up in a wealthy part of Oslo. Already at a young age, psychiatrists had recommended that he be removed from his single mother who was mentally unstable and said at times she wished he were dead. He had sporadic contact with his father, but this ceased when he was 16. He had few friends until he embraced and was embraced by the dark web in his 20s as he became radicalized. The gunman in the New Zealand attack was an Australian who lived for some years in the South Island. He was known by neighbors to be a bit of a recluse, who despite being a loner offered to mow the lawn for neighbors. His own father died in 2010, and from 2012, he had travelled widely in Europe and gradually radicalized himself, taking contact with far-right organizations and posting on social media platforms. He too was concerned with his notoriety and used social media to live stream his terrible deeds.

From Story 2, we see a number of points worthy of note: limited face-to-face networks and finding consolation on the Internet recalls the opening points on the changing sense of knowledge as networks of knowledge and connectivism seem ever-present. Adorno's point on replacement father figures is relevant to both gunmen and both shared the view that we were not stopping the arrival of immigrants. What we also note is the manner in which they hardened themselves to regard others as objects. They transformed their moral understandings and we are left with the last story to address the forward-looking point raised by Adorno - namely, how might we avoid through education that horrible events such as these occur again?

10.3 Story Number 3: Ressentiment, Moral and Values-Based Education

To say that the two gunmen were filled with anger and wanted to exert revenge would seem a rationale assumption, and also the task of education would be to prevent such emotions and beliefs from being formed through radicalization. Such a line of reasoning might be understood in one of two ways. It might be following Girard (1991) that what we desire is not determined by the object itself, as we normally believe, but by another person. He uses the example of fashion, where we desire a specific garment because we saw another with it on. For him, desire is mimetic, and in this context, the two gunmen were trying to emulate others before them whom they idolized.

While this theory might seem fitting, I would question if it is a good explanatory fit in this context. The gunmen may well have been inspired by others before them, but I would suggest that this offers at best, only a partial explanation for their actions. A second approach offers, in my opinion, an additional approach. I am thinking of the concept of *ressentiment* and what it can mean for a moral and values-based education. This is not to refer to the normal understanding of resentment, it is to allude directly to Nietzsche's understanding, whereby if a person feels wronged, they can do one of two things, either live out the revenge immediately in a spontaneous form of expression summed up as:

> "To be incapable of taking one's enemies, one's accidents, even one's misdeeds seriously for very long - that is the sign of strong, full natures in whom there is an excess of the power to form, to mold, to recuperate and to forget." (Nietzsche, 1969: 39)

Alternatively, they can cultivate an emotion of *ressentiment*. This is a reactive holding-in, a planning of revenge that can eat the person up over time and change them. To each of these actions, spontaneous revenge and *ressentiment* is allotted a moral way of thinking. With the former, it is a morality of good and bad, where the one acting out the revenge deems themselves as the good and all others are simply bad and worthy of no further consideration after the act. With the latter, it is the morality of good and evil, where the person wronged regards themselves as the good and the object is the evil one. The important point is the spontaneity versus the cultivation of the revenge, which may never actually take place.

In my work with Halland (Dobson and Halland, 1995), we proposed a pedagogy where the emotion and planning of *ressentiment* and spontaneous revenge is considered an expression of the *unsociable-social*, and it is important that children learn how to recognize and cope with both. To try and remove them is unrealistic. In a controlled classroom or community environment through role play or the like, children can develop such skills.

This is, however, only one part of the story. We also proposed that children learn the sociable-social non-ressentiment and showing no spontaneous revenge. This is the existential being with (*Mitsein*) others without actions or planned and felt ressentiment (Heidegger, 1962). It is to recognize the other as being part of our world.

I will close this story with three culturally contextualized examples of the sociable-social and again its central importance in a pedagogy seeking to address the question, what is the meaning of education? The point is to demonstrate that there are educational cross-cultural practices of the sociable-social and the list is much longer.

In Norway, there is a concept that refers to the importance of being gradually included into the mores of society through an education that is not simply about the content of disciplinary knowledge. It permeates all education, formal, informal, and non-formal, including morality and the values of society. The term is *dannelse* and the closest translation is the word *bildung*, meaning to be a well-formed identity able to be with others in a caring manner, without over-caring and taking from the other, a sense of responsibility and independence. Key, a well-known Scandinavian educator at the turn of the 20th century, expressed this and alluded to the taken for granted and hard to define, what in modern organizational science we might simplistically call soft skills, "bildung is what remains after we have forgotten everything we have learnt" [4] (Steinsholt and Dobson, 2011: 5).

---

[4]In the original Swedish: *Bildning är hvad vi hafva kvar, när vi glömt allt hva vi lärt – dannelse er det vi står igjen med når vi har glemt alt vi har lært.*

The second example concerns the concept of تَرْبِيَة (*tarbiya*). If dannelse/bildung is a secular concept, *tarbiya* is central to the Muslim faith and world view and infuses the lifelong education of each individual, inside and outside the formal school setting. It concerns the training of the soul through an Islamic moral education, where earning the praise of Allah is based upon surviving in a world designed to test and develop us. We are to become what we repeatedly do, and the act becomes the habit. The example might be to avoid becoming the donkey carrying books, oblivious to their meaning, transporting without understanding.[5]

The third example is about the relationship between Māori grandparents and their grandchildren as a lived everyday embodiment of the sociable-social. My poem expresses this where the shared root of the word for grandparent and grandchild is *puna*, meaning a spring - in this context, the spring of knowledge from the past and yet to be realized in the future of the coming generation.

Ocean

between tūpuna and mokopuna
on the porch always to be repainted
peeling a story silver feathered silence
strong backed and stubborn
carried by the whistling spring to Ocean
our Ocean
the sail cloth stained red
red the fish heads thrown back
to bait the dreams of other childhoods
of eyes turned down and precious thoughts kept for another day
of arms moved inward in sleep
our shelter

10.4 Conclusion

Through three stories, I have sought to explore and answer what is the meaning of education in a time of *ressentiment* and global hatred. The conclusion is that a moral and value-based education remains a central task and should

---

[5] Quran, Sura: 62, Verse: 4.

permeate all formal, informal, and non-formal education in school and other settings. This moral and value-based education should offer opportunities to learn about three components: *ressentiment*, spontaneous revenge and forms of bildung, tarbiya, and intergeneration communication — with deep respect for the past and the future yet to be with us. A central point made in this chapter is that merely using a teacher-based transmission model of pedagogy or a constructivism co-learning model of pedagogy is not enough. We must, as the examples of Norway and New Zealand show, engage with a model of pedagogy informed by connectivism and equipping all with the skills to connect different sources and networks of knowledge.

References

Adorno, T. (2006) Education after Auschwitz. In, Davies M. (ed.) In: How the Holocaust Looks Now. London: Palgrave Macmillan.

Aristotle (1965). *On the Art of Poetry*. London: Penguin.

Boje, D. (2001). *Narrative Methods for Organizational and Communication Research*. London: Sage.

Dobson, S. (2004). *Cultures of Exile and the Experience of Refugeeness*. Bern: Peter Lang.

Dobson, S. and Haaland, Ø. (1995) *The Pedagogics of Ressentiment. The Experience of Hamlet* Lillehammer: Lillehammer University College Press.

Girard, R. (1991) *A Theatre of Envy*. Oxford: Oxford University Press.

Heidegger, M. (1962). *Being and Time*. Oxford: Blackwell.

Nietzsche, F (1969) *On the Genealogy of Morals*. New York: Vintage Books,

O'Malley, V. (2019). *New Zealand Wars. Ngā Pakanga o Aotearoa*. Wellington: Bridget Williams Books

Ricoeur, P. (1984). *Time and Narrative*. Chicago: University of Chicago

Siemens, G. (2005). Connectivism: A learning theory for the digital age. In, *International Journal of Instructional Technology and Distance Learning*, vol. 2, pp3–10.

Seierstad, Å. (2019). The Anatomy of White Terror. In, *New York Times*, 18th March.

Steinsholt, K. and Dobson, S. (2011) (edited). *Bildung. Introduction to an opaque educational landscape*. (published in Norwegian: Dannelse. Utsikt over en ullendt pedagogisk landskap). Trondheim: Akademika Publishers.

11

The Interreligious Dialogue as a Premise to the Culture of Peace

Prof. Roberta Santoro

University of Bari Aldo Moro, Italy

11.1 Migration Flows and Cohabitation in Multicultural Contexts: The Value of Diversity

The current migration flows and their "modern" manifestations need a revision of the interpretative categories in order to better understand the dynamics of the phenomenon. It is known that the term immigration means "permanent or temporary movement of groups from one territory to another one, from one location to another one, determined by various reasons, but essentially by necessities of life. Migration can be "mass migration" or "infiltration migration", depending by the fact that they occur for a large amount of people (in this case they are mainly permanent) or for small contingents, so that in the new territory towards migration has taken place the ethnic group is not radically amended."[1]

Generally, it is a phenomenon, in our biological or social case, in which there is a movement of individuals, mainly in groups, from one geographical area to another, determined by environmental, demographic, physiological, political, social, and economic changes.

Among all the demographic phenomena, migration is the most unpredictable. It is an evolutionary process involving different adaptations over time, in which three main actors act: migrant individuals, the society of origin, and the host society.[2]

---

[1] Immigraziones (eng: immigration) in www.treccani.it

[2] Look at the annual report of the United Nations Department of Economic and Social Affairs (UNDESA). The number of migrants in the world in 2019 is estimated at 272 million,

Actually, there is no uniform legal definition at international level of the term "migrant," commonly used in a generic form with reference to both migrants and refugees[3] — and not rarely migrants and refugees coincide, especially because current contexts show a map where armed conflicts are more and more widespread.

In this perspective, the current migration phenomena are a modern and current manifestation of significant impact and also because it contains multicultural, multinational, multiethnic, or multiconfessional experiences and signs.

The legal category of multiculturalism, therefore, requires a redefinition of ways, functions, and rights within the same political society. This is not a phenomenon connected with the pluralism of interests, of individual needs, but with the cohabitation of cultures which ascribe meaning to choices and life plans of individuals within a more or less defined space.

In this sense, the notion acquires a polysemic character, from time to time characterized by legal, cultural, religious, political, and sociological elements.[4]

The data, according to which the transformed function of law in society enhances its instrumental value, is not insignificant because it is precisely through law that the relationships created between the various actors and between the different cultures are regulated — also taking into account all the conflicting aspects.

---

continuously increasing compared to past years. The increase of 51 million compared to 2010 data indicates that the number of migrants is growing at a higher rate than that reported to the entire world population.Migrants represent 3.5% of the global population, compared to 2.8% in 2000. The number of refugees or asylum seekers grew by around 13 million between 2010 and 2017. With 82 million, Europe turns out to be the continent hosting the largest number of migrants, followed by North America (59 million), North Africa, and western Asia (49 million). Nationally, approximately 20% of global migrants are hosted in the United States, with 51 million people. Germany and Saudi Arabia rank second and third, respectively."These data are critical for understanding the important role of migrants and migration in the development of both countries of origin and destination. Facilitating orderly, safe, regular and responsible migration and mobility of people will contribute much to achieving the Sustainable Development Goals" — said Liu Zhenmin, UNDESA Undersecretary.

[3]Migrants are a complex category from which we can be distinguish: refugees in the strict sense (Geneva Convention, 1951); people received under temporary protection; people received under humanitarian protection; people in protracted refugee situations; people displaced due to development projects, environmental crises, natural disasters, etc.

[4]The legal definition considers the proper angle of observation in a consistent manner with the discipline chosen as primary or as a simple starting point.

In this perspective, the phenomenon of multiculturalism applied to migratory flows generates "new" legal problems, such as the one concerning cultural rights, which justify their existence within the regulatory framework of a democratic structure both nationally and internationally for the very fact of deriving from a socially relevant phenomenon and closely connected with the global and legal development of society.

The character of modernity of multiculturalism lies in the fact that it tends not only to register diversity but also to govern it as a value in itself, looking at social dynamics, through intercultural processes.

In this dimension are highlighted important elements such as the recognition of otherness and the different traditional heritage, which belongs to the communities.

The multicultural society, which differs from the multiethnic one, especially in contexts showing migrant communities, must be able to prepare the necessary tools to ensure widespread development and the balance of cohabitative interests.

The true awareness of cultural identity and of the characteristics of diversity, constituting the unique specificity of the communities to which they belong, constitutes the necessary premise to experience the change within social systems and to understand the evolution of social, political, economic, and legal dynamics with logical connection and with compassionate moves.

It is worth noticing that, in a multicultural system characterized by the organized presence of migrant communities, the development of the social system is mainly a matter of ethical rules, legal norms, and fundamental values to be set and proposed as rules shared by the various communities.

In fact, the greatest difficulty encountered within our modern multicultural societies consists in the way of regulating and governing processes in an intercultural logic, through policies for regulating interests and through convergence toward shared objectives.

Therefore, multiculturalism and interculturalism are contiguous phenomena, necessarily integrated for a sustainable social development and for integration processes. It is not possible to imagine creating agreement toward objectives of composition of different interests if it is not allowed to all subjects — both physical and juridical — who live in a specific space to participate in the elaboration of consent and in the relative decision-making processes.

The multicultural society has to find its own expressive force, an adequate impulse, exactly from the existence of diversity, which, however, must be governed in order to turn the multicultural phenomenon into intercultural processes with which to encourage integration.

Governance must consist of the development of sustainable integration, dialogue, and knowledge, and policies in defined territorial contexts. Good governance is necessary to support democracy, to fulfill the protection of human rights, and to support social cohesion through solidarity paths, but, above all, it poses to political actors the problem of greater attention to the centrality of the human person and his existence issues.

In broad terms, it can be said that the various concepts used to describe the relationships between foreigners and the host society can be summarily divided into "integrative" and "disintegrative processes," depending on whether you want to focus on the inclusion dimension of the immigrant or on his exclusion and on the possible conflict between different ethnic groups.

That is why we are witnessing a constantly evolving change in the demographic composition of the EU, putting issues of control over entrances at the center of the political agenda and, at the same time, trying to adopt increasingly concrete interventions in order to achieve a better socio-economic insertion of the "guest" foreigner and a protection of personal rights, including the right to religious freedom.

The main commitment then becomes to trace a path toward integration that takes into account a real interaction of the different groups.

11.2 Multiculturalism and Interculture as Tools for the Composition of Conflicts and for Building Paths of Peace and Tolerance

Multicultural society, which is becoming more and more multiethnic, demands to be ensured as possible the recognition and sharing of a minimum nucleus of principles and rules that can form the basis for a common coexistence; this refers to the legal principle of equality before the law and the rights contained, in the first place, in the "Universal Declaration of Human Rights" and in the other international charters.

It is, therefore, necessary to identify a shared common base that makes it possible to pursue the construction of new systems of coexistence.

Our society requires dialogue, comparison, and discussion. Only the implementation of an intercultural education, which is based on respect for all cultures and on the recognition of equal rights and dignity — according to the principles of democratic coexistence — can give birth to this new type of community.

The end of 20th century was characterized not only by the collapse of the great ideologies and totalitarianisms but also by the rediscovery of the plural concept of democracy as a modern principle of community governance. Europe has gradually built up the principles and legal rules to regulate the peace and security process.

We need to only think of Articles 2 and 6 of the Treaty of Amsterdam on European Union, which define the common objectives to be pursued and the values on which to establish the common European coexistence. Evidently, this is still an ongoing legal–political process, in which the political will to contribute to peace, security, justice, and cooperation in Europe and in the Mediterranean, meeting some difficulties and obstacles.

The redaction of numerous legal acts — Helsinki Final Act (1975); Barcelona Declaration (1995); Treaty of Amsterdam (1997), in particular Articles 17 and 25; Treaty of Nice (2000), e.g., Article 17; the recent Treaty establishing a Constitution for Europe (2004) – undoubtedly highlights the importance of the final target, which is to make a coexistence in peace and security, promoting the development of peoples and their well-being.

The principle of everyone's participation in the governance of the society in which they live and with whom to establish particular bonds of belonging, definable as citizenship,[5] is connatural to the concept of democracy.

Certainly, not all relationships between people and communities within multiethnic and/or multicultural societies are suitable for achieving the legal value of citizenship, but only those formally identified by the legal system are suitable. However, the legal system registers and, somehow, juridically models social phenomena as an expression of social life and, consequently, regulates the intersubjective relationships internally, in proportion to the capacity with which it manages to look at problems by offering answers, beyond appearances and diversities.

It is, therefore, necessary to identify a new, legally relevant, broader, and more suitable concept of belonging for a new type of society.

In other words, it is necessary to re-elaborate a new citizens' charter, in which to regulate — with a different meaning — the framework of participatory democracy.

---

[5]Cf., Costa, Pietro. 2005. *Cittadinanza* (Citizenship), Bari, Laterza, 142–149; also, cf., La Torre, Massimo, 2004. *Cittadinanza e ordine politico* (Citizenship and political order), Torino, Giappichelli.

Moreover, united Europe needs to rely on a constructive social life and lay the groundwork for it, being capable of safeguarding real intercultural communication, giving, at the same time, concrete answers to arisen problems concerning conciliation of shared membership and diversity of life (as it results abstractly in the European Treaties and especially in the Treaty of Amsterdam, which gave birth to the European Union).

The Council of Europe, in its official documents, recognizes diversity as a source of mutual respect and social enrichment, also with the aim of thinking about citizenship — among the fundamental human rights — as expression of democratic culture.

Generally, citizenship, especially the active citizenship (which is related), represents one of the objectives of pluralist democracy, as already highlighted in European documents.

In relation to this objective, the prospect of active citizenship lies between the continuity of a pluralist and representative democracy and the strengthening of its participatory dimension, with the consequent possibility of experimenting with different forms and levels of social cohesion, declaration of personal and community dignity, growth of democratic culture, and exercise of responsibilities.[6]

This is an issue that goes beyond the simple consideration of rights and responsibilities as they are established in a juridical–formal dimension.

Basically, it is also an issue of educational policy encouraging and supporting joint participation for a new culture of democracy, as emerging from a broad overview of relationships among individuals, groups, associations, organizations, and communities, in which every citizen is actively engaged in cultivating solidarity values, increasing knowledge and attitudes for himself and in interaction with the whole of society. Hence, there is a need to seek new balances, new ways to coexist in societies, new relations between States (starting with an enlarged Europe), which have to strengthen processes and intercultural instruments, since they can no longer ignore their multicultural context.

The set of multiculturalism and intercultural aspects generates a necessary method in order to give new essence to coexistence, by which it could be possible to strengthen the values on which the work of building the international community and the European Union is carried out.

The coexistence of very different global cultural systems (since they stand on a different and non-homologous *humus*) — from an ethnic, religious, and

---

[6]Cf., Definitive text adopted by the Convention for young people, July 2002.

cultural point of view — must be organized in order to facilitate integration processes and an intercultural system, which is able to produce rules and values (a real right).

In particular, through the enlargement process after the end of bipolarism, Europe has seen the introduction of new multicultural models with the entry of new states — i.e., the introduction of visions of life inspired by a different humus, that is, multiculturalism, although to a lesser extent.

In this perspective, it is necessary to examine the contents of the common fundamental values of the Union, such as those indicated in Article 2 of the Constitution Treaty (human dignity, democracy, equality, freedom, rule of law, human rights, and rights of minorities), as the multicultural model is the best placed to guarantee value to national identities, as required by Article 2 of the consolidated version of the Treaty on European Union.

The process of transforming Europe into a large single community, as provided for by the Maastricht Treaty, 1992, and into a union of peoples (Treaty of Amsterdam, 1997) involves strengthening democracy also in regard to the advancement of new requests by new EU accession States and by States which have not yet entered into the Union but are willing to.

All the individuals making up people and communities, among which religions play an important role in building a shared area of peace, development, and security, contribute to strengthen democracy.

This is the meaning of the Euro-Mediterranean Partnership, set in motion by the Final Declaration of the Barcelona Conference (1995; involving not only each State of the European Union but also the States of the southern shore of the Mediterranean, as well as Cyprus and Malta), but which has unfortunately been abandoned following the events that shocked the Mediterranean.

It would be necessary to rethink the dimensions of the partnership set in the Conference, finding a way to rewrite the issues about security, fulfilling acceptable standards of peace, justice, and economic collaboration as well as building a humanitarian dimension, destined for cultural growth and human relations between peoples.

Compared to the new political and social scenarios and with a view to a different growth of social systems, the European Union finds itself in the condition of dictating the new rules in order to pursue a policy of peace and coexistence, which can support development.

> Developing mutual trust, a policy of peace and détente, using peaceful means for the conflicts resolution, connecting cooperation

to dialogue between peoples and, above all, creating conditions of economic justice.

These objectives can be achieved with the commitment of all social, political, and religious persons. When conflicts persist, religions create forms of collaboration, which begins with material help to the populations suffering due to wars. The collaboration regarding concrete affairs overcomes the theological difficulties and facilitates the rediscovery of the value of the authentic message of each religion.

At the same time, politics is required to recover its priority role also with respect to the economy itself in order to achieve the best protection of the human persons and their rights. In fact, the establishment of peace and the implementation of a security policy are the main way to realize the protection of the human person.

11.3 Building Peace with Dialogue and Integration Policies: The Role of Religions

Multicultural characterization, therefore, requires European culture to seek new ways to place its centrality.

The construction of a new humanism is affirmed in order to guarantee the cohabitation of differences in the same political, religious, and social space.

A bad management of the phenomenon of migration inevitably involves the non-acceptance of what is "different," denying openings to the "new" and hindering any form of encounter. All this leads to attitudes that concretize themselves — in the name of false foundations and prejudices — in a contrast to the social "pillars" and to those values set as essential conditions for the construction of a society truly based on the fundamental rights of the human being, i.e., on the concept of *humanitas*.

The very conflicting aspects — originating from the encounter and clash between different cultures — shape the conceptual structures and regulatory models of the various legal systems, characterize the interpretative categories of intersubjective relationships, and convey to common points and rules in order to regulate coexistence and cohabitation in the perspective of a peaceful conflict resolution.[7]

---

[7]Cf., Garelli, Franco, *La nuova centralità della religione nella sfera privata* (the new centrality of religion in the private sphere) in Burgalassi, Silvano, and Guizzardi, Gustavo (edited by), *Il fattore religione nella società contemporanea* (the religion factor in contemporary society), Angeli, Franco, 1983, 202.The author gives support to the idea that only if ethnic,

In its new enlarged geographical composition, the European space, on the one hand, had to verify "the existence of a phenomenon of exaltation" of diversity, and, on the other hand, it had to face the problem of relations inherent in religion and politics, within the relationship between the religious dimension and the law.

The main feature of our contemporary societies concerns "complexity," which shows itself as a disarticulation of structures and operating mechanisms of both individual subsystems and their mutual relationships.

The affirmation of religions, as one of the main factors of aggregation and cultural and social identity on a personal and collective level,[8] cannot be ignored by the welfare state in the provision of its services and in the determination of public policies.

The institutional response, at European level, has not always proved to be prompt and adequate; it has often been conditioned by security needs — the result of the historical context in which to define the rules of peaceful coexistence and security, as an element of development of people and their well-being.

The process of building peace between religions is part of the construction of Europe based on legal principles and rules governing the peace and security process.[9]

This process, which is a fundamental element for the whole Euro-Mediterranean area, can be supported with specific actions, such as dialogue,

---

religious, and cultural identities manage to be respected as such, it will be possible to build a multiethnic and multicultural society with the need to be able to rebuild a system of values which leads to peaceful coexistence, within which to be able to achieve the well-being of every man, as an integral part of society itself.

[8]Today, churches and religions are facing challenges never experienced before. Migration "is a phenomenon which impresses because of the number of people involved, because of the social, economic, political, cultural and religious problems it raises, because of the dramatic challenges it poses to communities, both national and international ... The phenomenon, as is known, is of complex management" – Benedict XVI, *Caritas in veritate. Lettera Enciclica sulla globalizzazione.* (Caritas in veritate. Encyclical letter on globalization.) Libreria Editrice Vaticana, 2009, 62.

[9]See, in this connection, Article 2 and 6 of the Treaty of Amsterdam, where the common objectives and the values on which to base the common European coexistence are defined. It is an ongoing long legal–political process, which encounters considerable difficulties. The production of the numerous legal acts highlights the importance of the ultimate goal, that is, to achieve a coexistence of peace and security, also promoting the development of people and their well-being. In addition to Articles 2 and 6 of the Treaty of Amsterdam are also to be mentioned the following: Treaty of Nice (2000); Helsinki Final Act (1975); Barcelona Declaration (1995); Treaty Establishing a Constitution for Europe (2004).

which specifically represents an instrument of both social and legal relevance. Legal culture has led to consider it as a value for social stability.

Dialogue is indicated in the Treaty of Amsterdam as a fundamental element of the European social model, acquiring full legal recognition in the Treaties (Articles 151–156 of the Treaty on the Functioning of the European Union), which aims to elaborate European social policy and which is considered as one of the main instruments to promote economic growth, social cohesion, and environmental sustainability of the various development processes.

Because of its legal value, social dialogue is governed by concertation procedures, which involve all social partners in the protection of the persons concerned through discussions, consultations, and joint actions. The dialogue initiated in 1985, by the European Commission, has been characterized by a political and legal path, which recognizes its importance in compliance with the autonomy of the social partners, by way of the legal framework of the Lisbon Treaty.

The dialogue takes place, concretizes itself, and makes itself up not only through declarations and the so-called good policy but, above all, through dedicated actions and regulations, which aim to stabilize security and development of the human person.

The Declaration of Barcelona, 1995 — that is the founding act of a global partnership between the EU and 12 southern Mediterranean countries — in Section 11.4 looks at the interreligious dialogue as a very instrument of social, cultural, human partnership in order to promote mutual tolerance and basic cooperation, immediately eliminating prejudices, ignorance, and fanaticism. The human and cultural partnership, of which dialogue was a structural element, initially received little attention from the European and African institutional actors, which were more interested in the increase of economic and business activities.

Even if the Declaration actually no longer generates rules and lawsuits, its content keeps its substantial importance and contains indications that can be taken into account, such as indications concerning the role of religions, which in the Final Declaration were identified as instruments to foster mutual trust and knowledge.

Only after 2011, following the change of strategy to face international terrorism, a more careful reflection developed in Europe about the importance of intercultural dialogue.

Gradually, dialogue becomes more and more an instrument used juridically as part of the legislative construction through the production of appropriate programs. In the face of the challenges and needs resulting from social and international coexistence, religions must become part of the democratic process, without forgetting and betraying the authenticity of their religious message and, at the same time, without conditioning or mortgaging the development of democracy.

For instance, in an increasingly plural and multicultural context, Christian religions have felt the need to review the relationship with civil society and its institutions, rediscovering the sense of the common good, the value of the political dimension, and the spirit of democracy, as they were places of coexistence and cohabitation of different subjects and different communities. All this has produced a direct effect on the religious freedom of individuals and religious confessions and also of "new religious confessions, highlighting previously unknown needs, which require new regulatory interventions" for a better protection of personal rights.

Peace, as a common and shared good, was put at the foundation of many international initiatives (UN, OSCE, Euro-Mediterranean Partnership, etc.) and the need to identify safe rules which are able to protect peace as a legal asset arose especially after 9/11.

The legal principles which inspire the right to peace and security must have an "ultra-state" dimension since the individual state cannot face the challenge of globalization on its own.

11.4 Religions in the European Context

The location of religious communities/affiliations and their activism, in a context which is both national and European, raises the issue of identifying common legal principles in the legal discipline about their relations with states. Furthermore, the need for religions to strengthen the process of collaboration and dialogue between them arises exactly from the search for this common discipline.

In order to establish compliance with what is enshrined by the law of the individual Member States' legal systems for what concerns the legal status of churches, associations, or religious communities (which are equated to the "philosophical and non-confessional organizations"), Article 17 of the Treaty on the Functioning of the European Union — which comes from the Treaty of Amsterdam — recognizes the specific contribution of these same organizations also through the activation of an open, transparent, and regular dialogue with churches and organizations; a dialogue to which the Union itself undertakes. Once again, the path of collaboration and dialogue, which has already been experienced in many of the States of the European Union and has been accepted as an instrument in the Treaty of Amsterdam, is that undertaken by Europe recognizing the importance of the "specific contribution" that religious communities and affiliations can offer without renouncing its secular connotation.

This contribution is important in relation to the need to soften possible conflicting situations caused by the increase in religious inhomogeneity due to the substantial non-European and intra-European migration flows. Along these lines, the Recommendation of the European Parliament of 13 June 2013 to the Council for what concerns the draft of EU guidelines about promoting and protecting freedom of religion or belief, as referred to point (o) which states that "within the framework of the elaboration and the implementation of the guidelines, support and commitment to a wide range of civil society organizations, including human rights organizations and religious or belief groups, which is essential in order to promote and the protect freedom of religion or belief; therefore the human rights focal points of the EU delegations should keep regular contact with these organizations in order to promptly identify the problems that could arise in the context of freedom of religion or belief in the relevant countries."[10]

According to this — chronological order is important — there is the Report on EU Guidelines and the mandate of the EU Special Envoy for the promotion of freedom of religion or belief outside the European Union,[11] in which it is underlined in point 3, with regard to the EU guidelines of 24 June 2013 on the promotion and protection of freedom of religion or belief, which, in accordance with Article 21 TEU, the EU and the Member States are committed to promoting respect for human rights, such as principle that guides the EU's foreign policy; it strongly welcomes the fact that the 2013 EU's guidelines integrate the promotion and protection of freedom of religion or belief in the EU's foreign policy and external actions and calls, in this context, for further activities to be strengthened in order to raise awareness and to implement the guidelines.

---

[10]Cf., European Parliament recommendation to the Council (13 June 2013) on EU drafting guidelines about promotion and protection of freedom of religion or belief (201372082 (INI)).

[11]Cf., Report on EU Guidelines and the mandate of the EU Special Envoy for the promotion of freedom of religion or belief outside the European Union (2018/2155(INI)).

As early as the 1970s, the Holy See gave birth to diplomatic relations first with the European Community, then with the European Union, guaranteeing an apostolic nuncio in order to follow the sessions of the European Parliament, to represent the opinion of the governing body of the Catholic Church for both the elaboration of the most important documents and the international events in which the European Union is involved.

It is worth mentioning that precisely at European level, the Church started in the 1970s a process of total organizational restructuring. In 1971, in fact, the Council of European Bishops Conferences (CCEE) was born, which is a body serving the Bishops' Conferences of all Europe, with the aim of promoting collaboration between bishops in Europe.[12]

In 1980, the Commission of the Bishops' Conference of the European Community (CECE or COMECE) was set up, which is made up of bishops delegated by the national Bishops' Conferences of the countries of the European Union: an agency with a light structure, with a permanent secretariat in Brussels with the aim of promoting "a closer union and collaboration between episcopates and episcopates with the Holy See in matters which the European Community is interested in" (Article 3 of the Statute).

The same process of organizational innovation has taken place in Europe within the inter-ecclesial relations of the other non-Catholic, Protestant, and Orthodox confessions.

In fact, the Bishops' Conferences of the countries of the European Union are represented in Brussels by COMECE, while the European network of Protestant, Anglican, and Orthodox Churches is represented at the Union by the CEC, Commission of the Church and Society. The two different Commissions collaborate with each other in the realization of a common project for a Christian Europe, so much that COMECE has a committee of experts which takes a stand on the measures of the Commission and the European Parliament in the proposal phase and during the redaction of a first draft of the provision to be adopted.

In case of issues deemed of significant interest for religious denominations and in order to contribute to the development of a common will, a consultation phase is initiated between the Community bodies concerned and the representatives of the CEC, which — as a side effect — allows the overcoming of fragility and inner rifts between the various confessions,

---

[12]In 1995, its current Statute was approved in which the members of the CCEE are expected to be the European Bishops' Conferences, represented by their respective Presidents. The Statute also considers the possibility that Bishops who are not members of an Episcopal Conference are full members of the CCEE anyway.

overcoming the controversies related to the pre-eminence role of the Catholic Church, with the sole purpose of being able to achieve positive results in the community.

Religious organizations, at this moment, not only realize a qualifying moment of the European process but also act in a concrete way, asking the European institutions to protect religious interests because these is an expression of values at the basis of civil coexistence.

In this perspective, it can be seen how religious organizations have strengthened their institutional presence within the European territory, placing themselves as privileged interlocutors in the construction of the new Europe.

It should not be forgotten that relations between the Churches are placed within ecumenical relations, attempting to foster a path toward shared theological values and ecclesial practices.

In this regard, it is necessary to mention the 2001 *Charta Oecumenica*, which highlights that: "the Churches promote a unification of the European continent. Unity cannot be achieved in a lasting form without common values."

In this sense, together with the ordinary areas, the CCEE operates in other fields that broaden the horizon of the topics covered and discussed in the various meetings organized throughout Europe. Particularly important is the one concerning youth ministry throughout the world, the dialogue between Christians and Muslims in Europe, the defense of religious freedom with the sole objective of supporting a society where justice, freedom, peace, and protection of environment dwell.

Particular attention is also paid to social and legal issues concerning bioethics, the Church–media relationship, and new technologies. For example, through the portal eurocathinfo.eu, the Church in Europe establishes an information network between the different Episcopal Conferences across the continent; in the same way, a portal for young people in order to access all the initiatives that the Church carries out in this area has been created. Non-secondary attention is paid to human beings to the protection of human rights within Europe and to their personal, spiritual, and social situation. One thinks of the issues related to migration and the problems related to the demographic collapse: to family, education, and culture of respect for life in order to defend it in all its phases (from birth to death), as well as interreligious relationships to succeed in promoting healthy coexistence in a pluralistic Europe (cf., 2001 CCEE Report).

Precisely on the occasion of the meeting of the bishops of the 20 Churches of the Mediterranean, (Bari, 19–23 February 2020) on the theme "Mediterranean frontier of peace," a Synod was held on the central themes of the Mediterranean in which it was highlighted that "the Christian communities do not stop building alternative ways of peace and testament of our Christian style of being within reality by placing the person as focus."[13] In the Mediterranean Sea, the Churches and people are facing very great challenges. Among them are the ones concerning interreligious dialogue and the challenge of welcoming migrants.

The meeting strengthened the ties between the Churches, which have committed to setting up interreligious committees in order to achieve true hospitality and dialogue with the aim of building a common path where we can grow in our areas a culture of peace and communion, one new style of dialogue, welcome, and support between communities. A new way of being Church.

---

[13]Cf., www.vaticannews.va: Final report of the Apostolic Administrator *sede vacante* of the Latin Patriarchate of Jerusalem.

12

The Dialogue Between Democracies: The Resolution of Conflicts and the Protection of the Human Rights

Gaetano Dammacco

University Of Bari Aldo Moro, Italy

12.1 The Mediterranean as "multi-space": multi-cultural, multi-religious, and multi-traditions

In the centuries-old history of the Mediterranean, the relationship of proximity has always been characterized by an alternation of peace and prosperity on one side and conflicts and wars on the other side. However, the crisis that characterizes the Mediterranean today is different because it presents elements of novelty and of different gravity. In some ways, even the current situation reflects the peculiarity of the Mediterranean, which presents itself as a space of contradictions, as a space of aspirations toward common vocations, as a place of cohabitation of the "multi" (multi-cultural, multi-religious, multi-traditions, etc.). The current geopolitical profile, therefore, tends to record diversities, now considered as an element of division, now as an element of integration: if governments and political logics use diversity as a place of conflict, people have demonstrated (and demonstrate) that the differences are beneficial. However, there is a problem of governing the differences and identifying mechanisms with which to govern them, recognizing that they constitute a value in itself and a wealth, also from an economic point of view for Europe itself.[1] Among the new elements, we

---

[1] According to Eurostat projections, the total population of the European Union with 28 countries is expected to grow from 508 million in 2015 to 520 million in 2080, with an increase of 2.3% (+12 million units). An insufficient increase to compensate for demographic deficit

can consider the following: an crisis is widespread and affects all the shores of the Mediterranean; migration flows are an issue that affects the whole Mediterranean, which has become a kind of long border; and a geopolitical change is taking place (favored by the rigidity of the European Union), which has the effect of introducing new players with hegemonic aspirations (Turkey, Russia, and China). Furthermore, Europe's substantial indifference toward a Mediterranean inclusion policy is a real novelty since, over the centuries, Europe, through the various kingdoms and states, has always played a leading role. Furthermore, three elements constitute a real novelty in the Mediterranean geopolitical panorama (destined to produce further consequences): the relevance that human rights are taking on especially in the perception of people; the value of the multi-dimensional dimension (more cultures, more religions, more traditions, etc. as an indispensable identifying element of people); and globalization as a category that enters each specific problem. Neighborhood relations (both for better and for worse) are governed by these elements of novelty that are grafted into the plural dimension of a Mediterranean, which continues to have a common vocation, a common destiny, and a common heritage.

---

and aging, two of the phenomena that risk compromising European economic well-being. To stabilize the population of the European Union around 500 million people, an annual flow is needed which increases from around 2.6 million foreign immigrants in 2015 to over 2.9 million in 2020. In 2030, if the projections will be confirmed, the average number of foreign immigrants will exceed 3.7 million. As regards Italy, more detailed demographic forecasts by ISTAT, limited to the population of working age (15–64 years), analyze the probable demand for foreign workers by the Italian production system, in a longer perspective, from 2015 to 2065, obviously gross of the inactive. In general, the central scenario of the forecasts of the Italian Statistical Institute foresees an increase in the overall population much more contained than that of Eurostat (the estimate of the total population in 2065 is 61.3 million; 65.8 million for Eurostat). The total working-age population will decrease from 39.8 million in 2015 to 33.5 million in 2065 (–6.3 million), as a result of a decrease of 10.4 million Italians and an increase of 4.1 million foreigners (the percentage of foreigners out of the total will increase from 11.3% to 25.6% in the same period). Consequently, the increase in immigrants will not be able to compensate for the reduction in the working-age population but will have unbalanced effects between the Center-North and the regions of the South. In fact, during 50 years, the increase of 4.1 million potentially active foreign workers will be absorbed by 4.1 million units from the northern regions, 1 million from the central regions, and only 400,000 from the southern regions. So, across Europe, the influx of emigrants is seen as a resource and as the solution to an economic-demographic problem. Obviously migratory flows, especially those coming from countries where wars or economic hardship, are stronger and with very high levels of illiteracy, will create serious problems as regards the difficulty between supply and demand for professional figures by companies in each country. European

All this makes the Mediterranean a fragile hinge between parallel and coexisting realities (made up of peoples, policies, cultures, religions, different traditions, which sell back a sort of primacy or, at least, of equal dignity): a "multi-verse" (not a "universe"), similar in some respects to the physical or philosophical conception,[2] destined to develop in a "finite" (and not infinite) space like the Mediterranean in which, therefore, a collision between the various elements is possible. The current critical issues are also an expression of a structural feature of the Mediterranean (i.e., that of being the place of diversity), which projects itself into the multi-verse supported by interests that come from far away from the Mediterranean. Just as the bipolar system changed the lives at Mediterranean people and countries until 1989, the current pushes are destined to bring about change, in an even more traumatic form. Globalization has opened the borders of the Mediterranean and has made them permeable to other universes (above all of oriental origin such as Russia and China), which have a different view of the facts, of the problems, and of the conflicts and offer different possible resolutions (which can also create new conflicts).

New phenomena have a profound effect on the dynamics of development, and, in a context of globalization, they take on a different, albeit contradictory,

---

[2]The term "multiverse" is adopted here only for the lexical ability to represent some geopolitical innovations. The term, coined in 1895 by the American writer and psychologist William James, in fact belongs to the scientific, philosophical, and theological world. From a scientific point of view, for the first time, Hugh Everett proposed it in 1957 with reference to quantum mechanics and was subsequently taken up by other scientific theories, on which we do not want to enter. The term multiverse, in reality, has an older philosophical meaning (Greek atomists believed a plurality of Earth-like worlds existed), which found a reworking in the Middle Ages, thanks to the studies of theologians and scientists such as the theologian Robert Grosseteste at the turn of the XII and XIII centuries, Nicolò Copernico between the XV and XVI centuries (with the discovery of the actual size of the universe, containing billions of galaxies), Nicolò Cusano in the XV century, and Giordano Bruno in the XVI century. At the basis of the thesis of the multi-verso, in many aspects not easily demonstrable, there are questions of religious content to which the great religions try to give answers (in Judaism, Hinduism, Christianity, Islam, New Age, etc.). The volume of Mary Jane Rubenstein, Worlds without End. The Many Lives of the Multiverse, Columbia University Press, 2014 is also interesting for the multidisciplinary approach, although not entirely acceptable in some analyses. Franco Cassano's analysis is of particular interest, who affirms that the Mediterranean "is not a monolithic identity, but a multiverse that trains the mind to complexity," a thesis developed in various publications also aimed at new forms of dialogue, cf. Danilo Zolo and Franco Cassano, The Mediterranean Alternative, Feltrinelli, 2007.

importance since if, on the one hand, there is an effort to support in any way the fundamental rights of freedom and of the human person, from the another even more sophisticated and violent is the attack on the protection of the fundamental rights of the human person.

12.2 Plurality of Models of Democracy: Is There an Arab-Islamic Model?

The "multi-verse," which characterizes the Mediterranean in this historical phase, underlines for some aspects the intrinsically plural character of this space, which is unique and singular.[3] The dimension that the Mediterranean is assuming, which can be defined as multi-verse, does not cancel out its characteristics but enhances them, introducing a new cultural model that is slowly being built even if conflicts and major problems continue. Over the centuries, diversities have grown in the Mediterranean geo-historical space: they have learned to cohabit (now clashing, now meeting, now fighting, and now talking); they have created a plural identity as a set of differences. This legacy today meets a new challenge, which consists in the contextual presence in the Mediterranean of different "worlds," which have their origin outside the Mediterranean and are represented by subjects who claim a new historical role in the basin of the mare nostrum. These subjects (states, economic and commercial organizations, religions, and power groups) who propose themselves as direct actors, who no longer seek mediation, are the expression of a multi-verse based on rules, ethical, and religious principles, on traditions, and on political logics that come from territories and visions of life outside the Mediterranean.

The plural model of the Mediterranean, as an expression of a culture shared by peoples of all sides, today is insufficient to offer suitable answers for the realization of peace policies and for the resolution of conflicts. Furthermore, it should also be noted that, while a change of perspective is required in search of new models of coexistence, cultural patterns inherited from previous situations still exist, expressions of a relationship between the north Mediterranean and the south marked by a relationship of subjection (colonial, cultural, and political), which in the post-war period experienced various

---

[3]Fernand Braudel was certainly the most attentive among the numerous authors to the theme and the problems of the Mediterranean, numerous writings, which still retain a current importance and are therefore reprinted; cf., most recently in the Italian language edition Il Mediterraneo, Bompiani editions, 2017.

stages. Today there are still strong limitations on the point of democracy. We cannot ignore that the north shore countries have tried to export their model of democracy and have judged the validity of the political experience of the countries of the other shores (both the south and the east) in the light of the western categories. In addition, Western powers have also recently attempted to intervene by force, ignoring internal movements and the popular innovative drive — without there being any justification, as has been shown, if not that of the affirmation of a dominance dictated by geostrategic and economic interests.

The popular push occurred during that phenomenon called "Arab springs," initially founded on a popular uprising movement. The sentiments were precise: the fall of authoritarian regimes (which did not displease certain Western politics), the moral change of social life gradually away from religious values, and, consequently, the fight against corruption, social degradation, growing poverty, the lack of freedom, and the unequal distribution of the wealth of the countries. There was a widespread aspiration to build social systems that respected the fundamental rights of the human person (heavily violated) and the deeper values of religion (Islamic and even Christian). The social models desired by popular uprisings had to guarantee political freedom, democracy, and social justice. In the Arab region repression, tyranny, absence of rights and freedoms, and massive violations of human rights went hand-in-hand with the concentration of power in the hands of the restricted elite linked to the party or family of those who governed. However, after 10 years, the bitter observation remains that little has been done in the direction hoped for by the people. The problems concerning the social structure and the political structure of the various countries remain standing and, among these, certainly remains the problem of democracy, a fundamental legal principle for the establishment of a modern state to protect the human person and his fundamental rights.

The aspiration to create a democratic system for the protection of fundamental freedoms is strongly present in the Mediterranean countries of the south and east shores as demonstrated by the 2012 Al-Azhar Shaykh Declaration which reads: "the Egyptians and the umma Arab-Islamic, after the liberation revolutions that have freed the freedoms giving new impetus to the spirit of the Rebirth (nahda) in all areas, they turn to the 'ulamâ' of the umma and its intellectuals to define the relationship between the general principles of the noble Islamic sharî'a and the system of fundamental freedoms on which international treaties agree and from which the experience of civilization of the Egyptian people originated." The document is divided into four points (freedom of faith, freedom of opinion and expression, freedom of scientific research, and freedom of artistic and literary creation), which recall the principles of a democratic system and repropose the topicality of the debate on democracy in Islam. In Arab-Islamic culture, the themes of secularism and democracy take on a crucial importance because they call into question the delicate relationship between religion, at the center of the public and private sphere, and political power, which has its source in religion but which is essentially secular.

In this new scenario, the Mediterranean has to face a new challenge, which, at the same time, is a new vocation, that is, cohabitation and dialogue as a way to peace and progress. The presence of new subjects must not create new forms of submission, but their active presence must favor any form of dialogue since the interests of these new subjects consist in the pacification of the Mediterranean. Dialogue is an instrument that can only be affirmed between subjects who realize a democratic experience; this means that democracies, which face the Mediterranean, must organize the forms of dialogue. However, do the social and political experiences existing on the shores of the Mediterranean define the existence of democracies?

The Arab-Islamic path to democracy also raises two questions: is there compatibility between Islam and democracy? If so, what are the essential features of this democratic model? If we consider the demands of the "spring," we can observe the existence of a demand for democracy, that is, for freedom, for the dignity of the human person, and for respect for moral and religious rules. In Arab-Islamic culture, we do not find a concept of democracy coinciding with the canons that define democracy in Western culture. However, the essential elements of democracy exist in the Arab-Islamic culture and experience, where we find the values of popular sovereignty, participation, consensus, and decision making (values also present in the demands of the "springs"). The current "reformist" movement (Al-Islah) is not new because it existed since the first half of the 19th century in a very large Islamic territory that went from the Maghreb to India, animating a new class of officials and intellectuals, deeply attracted by the culture and organization of European societies and determined to propose a modernization of their administrative, institutional, scientific, and cultural system by looking at Western models, but without denying the Arab tradition[4] In Islamic culture and experience,

---

[4]Among the most illustrious are the Egyptians al-Gabarti (1753–1825) and al-Tahtawi (1801–1873), the Tunisian Hayr al-Din (1822–1890), and the Indian Wadi Allah (1703–1762). Toward the end of the 19th century, the intellectual movement entered a more delicate phase: some reformists (islahiyyun) began to research more deeply the internal causes that,

consensus (Jjmâ') is important, involving "a large group of people or the majority of the people"[5] and, therefore, is considered "the legitimate foundation of a true theory of popular sovereignty"[6]. The concept of consent must be considered together with the concept of consultation (shûrà), which takes on different organizational modules in the various countries (for example, in the Egyptian system, there is a Council of the Shûrà, Majlis al-Shûrā, which is an advisory parliamentary body and elective of which the Governorates are part: a sort of House of Representation). A debate has reopened around these concepts and these institutes, tending to recover the fundamental elements of a recognized Islamic democracy[7] since, albeit according to a complex and fluctuating path, Jjmâ 'e shûrà (i.e., consensus and consultation) represents "the cornerstones and roots to develop a free and representative form of Islamic government or democracy."[8]

---

over the centuries, had determined the scientific and cultural immobility of the Arab-Muslim world. Among all the most important intellectuals who promoted reformism were the Syrian al-Kawakibi (1850–1902), one of the first theorists of pan-Arabism, the Lebanese Sakib Arslan (1869–1939), the Persian and Shiite al-Afghans (1838–1897), a strong supporter of pan-Islamism, the Egyptian imam 'Abduh (1849–1905), a true theorist of the most mature age of reformism, the Syrian Riîà (1865–1935) founder of the Salafiyya movement, Ahmad Khan (1817–1889), of Persian-Afghan origin, promoter of a vast cultural rapprochement with the West and, finally, the Algerian Ben Badis (1889–1940), defined the "herald of Algerian Muslim reformism and Algerian nationalism" (see Paolo Luigi Branca, Voices of Modern Islam: Arab-Muslim thought Between Renewal and Tradition, preface by M. Borrmans, Marietti, Genoa 1991).

[5] Ahmad Moussalli, The Islamic Quest for Democracy, Pluralism, and Human Rights, Gainesville University Press of Florida, 2001.

[6] See Luca Ozzano, Islam and Democracy: Problems, Opportunities and Development Models, taken from the website.

[7] About the relationship between the profound crisis of Islamic political institutions and the difficult process of democratic transformation, see M. Campanini, Ideologia e politica nell'Islam. Fra utopia e Prassi, Il Mulino, 2008.

[8] About the intimate connection between consultation (Ijmâ', as a manifestation of a will) and advice (shûrà, also as the foundation of an organism) with the spirit of Islam, cf., L. Ozzano, cit. Today, the word "council" is widespread in many Islamic countries, being used also for "Parliament"; however, a restrictive interpretation of this concept, developed over the centuries by Islamic theorists, has become prevalent, meaning the shûra simply as a faculty of the sovereign, concerning not the entire population, but only a small body of advisers. However, it seems that this interpretation of the concept of "consultation" has not always been the preferred one: concretely, the Prophet would have used it more widely and, after his death, would have become "a symbol identified with political participation and legitimacy." On the evolution of the forms that define the relationship between governed and governed, cf., Bernard Lewis, The Political Language of Islam, Rome-Bari, 2005, 51 ss.

Therefore, it can be argued that Islam can interpret a possible model of democracy since the "minimum requirements" are present, which allow a "minimum definition of democracy ... that is, by attributing to the concept of democracy some specific characters on which we can all be agree" (Norberto Bobbio).[9] In essence, this means that the model does not replicate another existing model but must contain the essential elements that make it possible to identify a democracy.

Therefore, the value of democracy does not consist in the purity of an abstract model or in a "unique" model. Democracy is an "attribute" of society, that is, a "procedural model" that allows the participation of the people who elect their own rulers and who foresees a power that makes the necessary decisions after a free and majority discussion. If we accept the idea of a "minimum definition" of democracy, we can accept the existence of multiple models or systems of democracy, different in form but not in substance and, therefore, compatible with each other. The democratic system must be founded on supreme cultural, spiritual, and even religious values. In this sense, there is an Arab-Islamic path to democracy based on values also of religious derivation, which reaffirm the value of the fundamental rights of the human person and his dignity. The existence of multiple Mediterranean models of democracy introduces a new perspective of dialogue, that is, between political, institutional, corporate, and democratic models.

In this new scenario, the Mediterranean has to face a new challenge, almost a new vocation, that is, cohabitation and dialogue as a way to peace and progress. The presence of new subjects risks creating new forms of submission, or a sort of new colonialism, for countries on the south bank. However, their active presence and all forms of dialogue must be encouraged since the interests of these new subjects are linked to the pacification of the Mediterranean, to the creation of economic and political relations of peace. The war economy does not enrich the system (i.e., the whole of society) but only a few groups. Dialogue is a possible tool only between subjects who realize a democratic experience: this means that also the democracies, which face the Mediterranean, must organize the forms of dialogue.

---

[9]See Norberto Bobbio, from an interview by Diego Fusaro, on the website www.filosofico.net/bobbio

12.3 Religion and Laicity in the Euro-Mediterranean Area: The Value of Practical Experience

The theme of values on which democracy is based refers to the question of the relationship between religion and laicity. Laicity is a cultural value that belongs to the history of individual countries, and this has generated different ideas and experiences. Laicity as an "expression of the political-juridical vocabulary has a history of at least two centuries. It is not a dogma, but a story. In the nineteenth century, secularism arose, from the Enlightenment culture, as an ideal and fighting position, to affirm the emancipation of the state, of culture, of education, from the ancient régime, dominated by the Church" (A. Riccardi). The origin of laicity finds its source in the Gospel when the distinction between what belongs to God and what belongs to Caesar is highlighted as existential dimensions and separate institutions.[10] The countries of the north bank, such as Italy, France, and Spain, have been characterized by the great influence of the Enlightenment, an experience that has many merits but also many critical issues, of which we still feel the reflections in Western culture today. With reference to the question of laicity, the Enlightenment culture developed a secular conception of morality, released from religion, so that religion (with its revealed truths) and morals (as thought based on reason) develop a line of separation.

A different path has been taken in Islamic culture. The word "laicity" (*almaniyya* or *ilmaniya*) is a neologism that dates back to the 20th century (between the 19th and 20th centuries, the term *ladiniya* was adopted, i.e., non-religious, while in Morocco the term *laykia* is used) and which was probably coined by Egyptian Christians. Often the word is confused with the word "atheism," generating a "semantic abuse." The debate about the relationship between laicity, religion, and secularism is very important and committed; it originates different positions, among which very interesting is the position of Ali Abderraziq, according to which "it is not possible to assert that laicity is rejected tout court by Islam"[11] on the basis of the sources and

---

[10]An interesting analysis is in Ombretta Fumagalli Carulli, To Caesar What belongs to Caesar, to God What Belongs to God. Laicity of the State and Freedom of the Churches, Vita e Pensiero, Milan, 2006, which highlights the contribution of canonical doctrine about laicity, following three areas of research (the emergence of the secular idea in European history, the position of the Italian state, the new frontiers of Europe).

[11]The statement refers to the work of Al-Azhar Ali Abderraziq theologian, L'Islam et les fondements du pouvoir (ed. La Découverte 1994), published in 1925, quoted by Abdou Filali-Ansary in The contrast between laicity and Islam it does not exist, transcription of the speech given by the author at the round table organized by Reset Dialogues on Civilizations

of all the Quranic verses in which one can believe or assume that there is an allusion to politics and its ideal conduct. However, there is a strong idea that between laicity (and all that belongs, such as politics) and religion, there is "an irreducible dichotomy and tangibile" (Abdou Filali-Ansary) and that religion always prevails.

However, it should be noted that at the basis of this idea, there is also a misunderstanding to be clarified, which could derive from a just principle regarding the need to moralize political life. The relationship between the religious and political dimension, between the community of the faithful (the *umma*) and the state community, can be understood in the perspective of faith-oriented Muslim humanism, which requires continuous attention for the earthly interests and needs as an adhesion to the divine will.[12] The debate on laicity is within the broader problem of modernity in Islam, which follows sometimes conflicting paths and lines of thought. The humanistic vision in Islam is based both on the foundation of faith and on the recognition of human reason (reason and piety are the expression of humanity itself).[13] In this perspective, the problem of laicity, and of the separation between politics and religion, can be observed as a practical and not ideological or theological problem. The political construction of Muslim society does not originate in a philosophical change, but first of all in a practical reason, concerning the best way to organize the political life of the community of the faithful in history, in order to best represent the faithful obedience to law. All the problems linked to laicity fall within this pragmatic framework, not marginal or minor. Therefore, going beyond commonplaces and easy judgments, it can be said that the practical datum of experience becomes the place of obedience to the

---

and carried out in the context of the UNESCO World Philosophy Day (Rabat – Morocco, 16 November 2006). Abdou Filali-Ansary claims that Abderraziq, looking through all the Prophet's hadiths that could allude to politics, discovers "that in all those verses there was absolutely nothing that referred to a certain conduct which Muslims should adapt to, nor any institution or form of government that can be defined as an 'Islamic political system'." He — and this is perhaps his most important contribution, which we, through this continuous learning, are called to enhance — has advanced the idea that a fundamental distinction must be made between Islam and Muslims.

[12] Still scrolling the sayings of the Prophet, we read: "the best of you is not the one who rejects the ground for the divine or vice versa, but the one who takes from both of them." It is in the spirit of the Koranic vocation "not to forbid the goods permitted by God."

[13] This supremacy of knowledge and knowledge is present in some of the Prophet's speeches, in which we read "the reflection of an hour is better than the seventy-year service" and more "certainly the ink of the sages is more precious of the blood of the martyrs."

divine will and nothing prevents Islam from being compatible with laicity, with democracy, with humanism, and with modernity.

12.4 Dialogue Between Cultures and Religions Instrument to Strengthen Democracy and Create Tolerance

The aspiration for peaceful coexistence and the plural and multi-verse characteristic of the Mediterranean require that the social, political, and legal instruments are identified to achieve this goal and to protect the fundamental rights of the human person. Among the tools, dialogue, increasingly frequently indicated in international treaties and acts, has a peculiar effectiveness. The word dialogue (from the Greek *dià*, "through," and *logos*, "speech") has multiple meanings. In any case, it can be understood as a comparison between distant or opposite parts aimed at finding an agreement. The need for comparison, as a good coexistence practice, is typical of a society in which communication is a substantial element. However, to be effective, the dialogue must follow its own rules; it must have a procedure, a method, a goal, and a system of sanctions. Dialogue must play a strategic function aimed at identifying possible solutions between conflicting parties. The logic of the dialogue is contrary to the "reason of the fittest" since it does not favor overpowering, but a change necessary for the achievement of a shared goal and the protection of a superior common interest. Our societies are characterized by multi-culture and plurality (of religions, societies, economies, legal systems, and visions of life), phenomena that pose a problem of relations between different people and communities, with inevitable consequences in terms of understanding, communication, and cohabitation. However, coexistence (between different peoples, religions, and cultures), which also constitutes a significant effect of the globalization process, is, at the same time, an obligatory way for a coexistence of peace and progress, as an exaltation of fundamental human values and as an alternative to a perennial state of war and destructive conflict. Coexistence arises from the awareness that there is a condition of equality and substantial equality between the different visions of life, the different cultures, and the different religions, although there are many difficulties in affirming the equal condition. In this situation of potential conflict, dialogue becomes a suitable tool to overcome divergences toward common goals and to favor integration processes.

Obviously, the dialogue we are talking about is not only a good disposition of the soul but an instrument with political, social, and legal contents.

From a legal point of view, these new instruments push the right to produce and strengthen universally valid legal principles to govern the social effects of cohabitation of diversity. This means, first of all, favoring the democratic participation of "different" communities in the definition of the rules of cohabitation. In fact, the law cannot be limited only to registering diversity, but must govern it; that is, it must produce superior principles and a system of rules, which, within the polis, allow "different" people and communities to participate in democratic processes, to protect fundamental human rights, minority rights, and the right to identity.

The dialogue, as an effective legal instrument, already has its own regulatory framework of reference, at European level and in bilateral relations, including relations between states and religious communities. Without considering the 1995 Barcelona Declaration (which indicated the dialogue between religions and cultures for mutual knowledge and trust), which has lost its effectiveness, there are numerous international, bilateral, area acts that refer to the dialogue as a tool to build a stable relationship: for example, the European Neighborhood Policy (ENP), born in 2003, which proposes action plans to strengthen more concrete and targeted cooperation; the projects of the European Union for the Mediterranean (UPM), born in 2008, which proposes forms of *ad hoc* cooperation on concrete projects; the Treaty on the Functioning of the European Union (TFEU), in which the "social dialogue," governed by Articles 151–156, constitutes a fundamental element of the European social model with its own procedures aimed at drawing up European social policy; and the "competitive dialogue" envisaged by Directive no. 2004/18/EC on the award of contracts for the construction of complex public works or the provision of public services.

Another example of the legal validity of the dialogue is that indicated in the concordats between the Holy See and the states, aimed at resolving legal disputes (for example, in the Italian concordat of 1984, Article 14: in the concordat between Poland and the Holy See 1997, Articles 27 and 28). The bilateral agreements of the Holy See with states, which especially with the pontificate of John Paul II have favored the path of dialogue, tend to protect the person and his inviolable rights, while general agreements in international fora (such as the OSCE) also tend to develop international texts containing humanitarian protection, in accordance with international law and according to the international commitments made to date by States.

We must mention the Council of Europe strategy which attaches great importance to intercultural dialogue, as an appropriate instrument to promote human rights, democracy, and the rule of law while also strengthening social cohesion, peace, and stability. The White Paper on intercultural dialogue "Living together in equal dignity" was launched in this direction. It was launched by the Foreign Ministers of the Council of Europe during their 118th ministerial session in Strasbourg on 7 May 2008. It stems from the awareness that cultural diversity is an essential condition of today's human society. Therefore, intercultural dialogue must be promoted to make "diversity" a source of mutual wealth, which fosters understanding, reconciliation, and tolerance.

12.5 Dialogue Beyond Tolerance: The Role of Religions

Cooperation and dialogue between religions, in the context of the complexity of the Mediterranean, assumes strategic importance as a way of achieving peaceful coexistence and balanced development (personal and social). Dialogue between religions is a necessary tool for cooperation between cultures and societies and suitable for achieving harmony. The initiative of the United Nations General Assembly lies in this direction when, in 2010, it adopted Resolution 65/5 of the month of October on the basis of which the "inter-religious week of harmony" was established. This was the final result of a long journey on the path of dialogue between cultures and religions started in 1999 with resolution 53/243, concerning the program of action for a culture of peace. The resolution on the week of inter-religious harmony recognizes the high moral value of dialogue between religions and beliefs for peace, mutual knowledge, and tolerance, and also considers inter-religious dialogue and mutual understanding as important dimensions for a culture of peace.

The UN message is transparent: to urge religions and cultures to dialogue and, at the same time, constitute a deterrent signal and a contrasting action for those who use religions for the purpose of violence. In addition, there are dialogue and meeting practices carried out by religious bodies, associations, and agencies, such as those of the Holy See on the subject of inter-religious dialogue and ecumenism (especially through the activity of the Pontifical Council for Interreligious Dialogue), and meetings of San Egidio's community (which since 1987 has held meetings of prayer and culture as a continuation of the Assisi meeting wanted by Pope John Paul II in 1986), the Ecumenical Assemblies (just remember the recent third meeting held in Sibiu in Romania in September 2007) promoted by European Christian churches and by representative organizations such as CCEE (the Council of Episcopal Conferences of Europe) and CEC (the Ecumenical Council of Churches, which represents about 350 churches and religious denominations present in more than 110 countries worldwide). Among the most recent initiatives, the international Islamic conference held in Madrid in June 2008 promoted the Saudi king Abdallah with the participation of the representatives of the monotheistic Muslim, Christian, and Jewish religions and of the Holy See.

These initiatives are of great importance also because they tend to stabilize the dialogue, reversing the trend compared to a recent past; they are even more relevant because they start from the recognition of the value of diversity and the importance of the heritage belonging to individual religions, which are recognized as equally respected experiences and as a necessity for the construction of a coexistence of peace, development, and solidarity.

Particularly interesting is the International Conference on Interreligious Dialogue promoted by the European Parliament in 2015 entitled "The rise of religious radicalism and the role of inter-religious dialogue in the promotion of tolerance and respect for human dignity," very much participated by religions. The dialogue between the different European cultures and religions dates back to 1992, by Jacques Delors, then president of the European Commission, for the need to create an ethical and spiritual dimension for European unity, going beyond economic and legal issues. This initiative, proposed periodically, finds its legal formula in Article 17 of the Lisbon Treaty, which provides for an "open, transparent, and regular" dialogue between the Union's institutions and the churches present in Europe. On this path, the project "Global exchange of religion in society" has recently been launched for the year 2020 with the aim of promoting all forms of inclusion.

In the delicate current moment, religions are "towards new areas of existence," different from those culturally and theologically considered traditional, more tolerant, and more inclusive, and closer to people's needs. This also happens because the problems and requests of the faithful grow, which strengthen the sense of belonging to a reality that constitutes the sense of existence. Migratory flows are an element that increases the role of religions, which take on a new awareness of the value of religious heritage. The meeting of religions goes beyond tolerance, even if this is an important goal. Religious tolerance is the condition through which the different faiths and practices of one or more religions, other than that professed for the majority within a people or nation, are accepted or allowed. It is, therefore, a concessive act that precedes any form of acceptance and makes sense if the ultimate goal of religions is to work toward common goals. In this direction is the subscription "Document on human brotherhood for world peace and common coexistence," signed on the occasion of Pope Francis' visit to the United Arab

Emirates in February 2019, signed by the Pope and by the Grand Imam of Al-Azhar Ahmad Al-Tayyeb. The objective of the Document, which underlines the importance of the role of religions in the construction of world peace (also highlighted in various international documents), is clearly outlined in the conclusions and can be represented as an effort to converge the two religions (Islam and Christianity) toward common goals founded on faith in God, namely: an invitation to reconciliation and brotherhood among all (believers, non-believers, and people of goodwill); an appeal to every conscience to repudiate violence and blind extremism; and an appeal to those who love the values of tolerance and brotherhood, promoted and encouraged by religions.

Tolerance is only the first step because the way in which religions can be faithful to their message and to the service of the faithful and of people is the search for a common commitment to achieve common goals, in other words, to practice faith together and each with faithfulness to God.

13

Tolerance and Nonviolent Practices

Prof. Hugh J. Curran

Peace and Reconciliation Studies, University of Maine

In the following essay, I will refer to Celtic spiritual traditions: Tolstoy's "The Kingdom of God is Within You," and Gandhi's biographies including the "Eighteen Last Years," each with their focus on nonviolence and tolerance.

The etymological meaning of "tolerance" is "to endure or support" and derived from it is the word "extol" meaning "to exalt" or "raise high," whereas violence is derived from "Violare" which means "to force." A Sanskrit word for Violence is "himsa" and its contrary is Ahimsa (Nonviolence), an ethical term suggesting an underlying attitude based on good intentions, good thought, and active engagement with issues. Nonviolence tends to be pro-active, whereas passive resistance tends to be reactive.

In terms of Celtic spiritual traditions, the period from the 6th to the 11th century placed an emphasis on spiritual practices involving nonviolence and tolerance. Chanting the following Beatitudes from the "Sermon on the Mount" was a daily practice in many Celtic monasteries: "Blessed are the poor [humble] in spirit for theirs is the kingdom of God; Blessed are those who mourn for they shall be comforted; Blessed are those who hunger and thirst for righteousness; Blessed are the merciful. . .; Blessed are the pure in heart for they shall see God; Blessed are the Peacemakers for they shall be called sons of God; The Golden Rule was often invoked, that one should "Do unto others as you would have them do unto you." Jesus' teaching requires both agape and righteousness. In both, attitude and action, inward and outward become one. . . listeners are summoned to an unlimited responsibility that they cannot escape." (1)

One of the outstanding clergies of the Celtic spiritual period was St. Eunan (aka Adomnan, 624–704) the Abbot of Iona, Scotland, who wrote a biography of his 6th century ancestor, St. Columba, the founder of Iona Monastery. Eunan began a campaign in the 7th century known as "The

Law of the Innocents" (Cain Eunan) or "Peace of Eunan." It was said that his supporters met with 100 chieftains throughout Ireland, Scotland, and Wales and convinced them not to involve women, children, the elderly, or clergy in warfare. This agreement, which held true for centuries, was a considerable accomplishment at a time when violent wars between clans was widespread. (2)

The 20th century Celtic scholar, Myles Dillon, noted similarities between the Celtic and Indian traditions based on the fact that they were located in the extreme ends of the Indo-European world and were conservative cultures that retained certain features in common. Dillon noted in his book Celts and Aryans that "the Indo-Europeans believed in Truth as the supreme power by which all creation is governed..." [It was] "a life-giving principle and sustaining power in the world which appeared in Greek thought in 500 BC as the "logos" (Truth) of Heraclitus, and later in the Gospel of John as "logos Kyriou." (3)

In 1880, Noncooperation was introduced in County Mayo in order to redress wrongs perpetrated by the landlord agent, Captain Boycott, who was demanding higher rents than poor tenants could afford to pay. This Noncooperation was successfully prosecuted with the help of the Land League, a newly formed national organization intended to help beleaguered tenants. This "boycotting" process was adopted by Gandhi as Noncooperation and became an essential part of the nonviolent movement for Indian Independence.

In 1920, fasting was invoked by Terence MacSwiney, Lord Major of Cork, who was imprisoned by the British during the Irish Independence movement. He undertook a hunger strike that lasted 74 days, resulting in his death on 25 October 1920 at the age of 41. His fast unto death gained worldwide attention with Nehru stating that he was inspired by his example, and Gandhi considering him a major influence on his life. (4)

More recently, in 1981, 10 hunger strikers fasted to death in the Maze Prison in Northern Ireland in order to draw attention to the need for better conditions in prison and also to protest the lack of civil rights for Catholic Nationalists. One of them, Bobby Sands, was elected an MP to the British Parliament shortly before dying of starvation. His death, with the other nine, galvanized anti-British feeling among the Irish diaspora. This led to more serious attempts at reconciliation and, finally, to the Good Friday Agreement on 10 April 1998. Soon afterward, British troops were removed from Northern Ireland and an open border with the Republic of Ireland became a normal part of cross-border activities. (5)

Halfway around the world, Britain's other colony, India, was also endeavoring to gain Independence and Mohandas Gandhi took up fasting as a strategic and spiritual practice. Following World War I, he also adopted nonviolent noncooperation against the British Raj even though not all of his fasts were political. His first fast in India was in support of workers who were refused a livable wage by mill owners in Ahmedabad. Gandhi's fast was decisive in a settlement agreement. The principles that he adopted were Satyagraha (Truth Force) and Nonviolence (Ahimsa), two principles rooted in Jain, Buddhist, and Hindu tradition.

Gandhi was initially inspired by Count Leo Tolstoy, whose book The Kingdom of God is Within You, had a powerful impact when he first encountered it in South Africa, where he had lived for 22 years. He founded Tolstoy Farm which was supported by a merchant benefactor and then brought his wife and family from India to join him. This spiritual community was inspired by the principles of Nonviolence and Truth that Count Leo Tolstoy promoted. (6)

Tolstoy's own spiritual journey toward Nonviolence and Truth seeking began after his participation as a soldier in the brutal war at Sevastopol. Following the Crimean War, he resigned from the military and had extended stays in Paris on two separate occasions (1858 and 1860) where he encountered Victor Hugo and other notable authors. One of them encouraged him to read a recently translated German version of The Tirukkural (Sacred Verses), a classic Tamil text dealing with: "Aram" (virtue), Porul (polity), and Inbam (love). In Aram, the text focused on moral vegetarianism and nonkilling (Ahimsa), basing itself on secular ethics that "expounded a universal, moral and practical attitude towards life... [and] spoke of the ways of cultivating one's mind to achieve other-worldly bliss in the present life...." Nonharming was considered the "foremost of virtues, even above [that of] Truth". A quote from the Tirukkural itself indicates the emphasis placed on Nonviolence: "Not killing What is the work of virtue? 'Not to kill'; For 'killing' leads to every work of ill. Never to destroy life is the sum of all virtuous conduct. The destruction of life leads to every evil. Let those that [are in] need partake [of] your meal; guard everything that lives; his the chief and sum of lore that hoarded wisdom gives. The chief of all (the virtues) which authors have summed up, is the partaking of food that has been shared with others, and the preservation of the manifold life of other creatures." (6)

Tolstoy was inspired by his readings of the Tirukkural and his discussion with other French thinkers, including Pierre-Joseph Proudhon, an anarchist writer who wrote "La Guerre et La paix" (War and Peace), a title Tolstoy borrowed for his own book. According to Tolstoy, Proudhon "was the only man who understood the significance of education and the printing press in our time." Returning to his estate in Yasnaya Polyana, he founded 13 schools for the children of Russian peasants, based on democratic principles. This was helped by the fact that emancipation from serfdom had taken place in 1861. (7)

After devoting himself to writing his renowned books Tolstoy went through a profound spiritual crisis and then wrote "The Death of Ivan Ilyich" and "What is to be Done" in 1886. He became known as a radical pacifist and devoted himself to exploring the implications of Nonviolence. His "Letters to a Hindu" in South Africa helped set Gandhi on a path that was as radical as Tolstoy's but with the need to assimilate European and Indian religious ideals and religious practices. Tolstoy's views on Truth and Nonviolence became core principles which encouraged Gandhi to explore Indian philosophy and religion in order to re-discover similar principles from his own tradition.

Tolstoy's reading of the "Sermon on the Mount" and the Golden Rule "that they should do unto others as I would that they should do unto me" became a deeply held belief, and even an obsession. He asserted, "only when I yield to the intuition of Love...is my own heart happy and at rest...The Divine work...which [can be] accomplished in this world and which I participate in by living, is comprehensible to me...[and] this is the annihilation of discord and strife among all men and among all creatures and the establishment of the highest unity, concord and love." As he noted in a letter at the time, "The way to do away with war is for those who do not want war, who regard participation in it as a sin, to refrain from fighting." (8)

After founding Tolstoy Farm, Gandhi agreed in principle with Tolstoy's intent to bring to fruition the "kingdom of God on earth" but formulated his own version as the "The kingdom of Rama on earth" which was to be accomplished by Nonviolence as the means and Truth as the end. He noted that, "The first step in Nonviolence is that we cultivate in our daily life...Truthfulness, humility, tolerance and loving kindness...[while] against violence [we must be] ever-wakeful, ever vigilant, ever-striving...." (9)

After returning to India Gandhi articulated this as, "Divinity is omnipresent and sits in the hearts of all...[although] we cannot grasp the essence, even an infinitesimal fraction, [but] when it becomes active within us, it can work wonders." He continued, "Ahimsa (Nonviolence) is [found] in Hinduism, it is in Christianity, it is in Islam...I have heard it from many Muslim friends that the Koran teaches the use of Nonviolence. It regards forbearance as superior to vengeance. The very word Islam means peace, which is nonviolence. Badshakhan, [Abdul Ghaffer Khan] a staunch Muslim never misses his Namaz and Ramzan." This allusion was to Badshakhan's involvement with Gandhi at his early morning meditations at his Ashram. Badshakhan would later lead Pathans in the border provinces in nonviolent protests based on Gandhian principles. His nonviolent actions led to incarceration in British and Pakistani prisons for over 30 years.

In Gandhi's Ashram, in addition to invoking Rama, Muslim prayers, as noted above, were offered and Christian hymns were sung. Gandhi stated that when he referred to Hinduism, he was including Jainism and Buddhism. His eclecticism was understandable given his background in London attending Christian services, while later in South Africa, he developed deep friendships with Indian Muslim merchants. The merging of prayers and invocations of all three religions in early morning devotions would lead him to insist on the sayings: "Truth is God; Love is God," as a way of transcending theological differences. (10)

After World War I Gandhi became more deeply infused with Indian aspirations for Independence from Britain and also became more immersed in the Bhagavad Gita (the Song of God), which encapsulated his strong faith that the Gita represented the eternal duel between forces of evil and good and that good would ultimately prevail. (11)

Although Gandhi's fasts were often strategically motivated, his strict vegetarianism and belief in extensive fasting were inspired by his mother whose devotional life had included frequent fasts associated with Jain asceticism. Prior to leaving for London to study law, Gandhi made a vow to his mother to maintain a strict vegetarian diet. The importance of vows and pledges became profoundly important methods he adopted in his organizing campaigns for redressing social wrongs in South Africa, and working to improve conditions for mill workers and poor farmers in India.

In an interview with Maurice Frydman in August 1935, he stated that "the only way to find God is to see Him in his creation and be one with [creation]. He continued, "I am part and parcel of the whole and cannot find [Divinity] apart from the rest of humanity." "Untouchability, khadi and village regeneration [is] where my energy is concentrated. Hindu-Muslim unity is my fourth love... I have never ceased to yearn after communal peace."(12)

In a later discussion with Frydman, he observed that "nonviolence implies complete self-purification" and "is without exception, superior to violence. The power at the disposal of the nonviolent person is always greater than if he was violent." (13)

Gandhi developed a plan for a "Peace Brigade" that would build tolerance. He insisted that: "a messenger of peace must have equal regard for all the principal religions of the world, which means they should possess a knowledge of different faiths." While intending to organize peace brigades, violence was beginning to take place on a massive scale when Europe was once again preparing for war.

On 6 October 6 1938, Gandhi stated that "the Peace of Munich" is the triumph of violence [but] it is also its defeat…[and asked rhetorically whether] "Germany or Italy added anything to the moral wealth of mankind?" (14)

When World War II was underway, India was asked to join the war effort in support of Britain. Gandhi wrote from his Ashram in Sevagram to the Viceroy, Lord Linlithgow, that it was the position of the Indian Congress that there was "…a conscientious objection to helping a war to which they were never invited and which they regarded…as one for saving imperialism, of which India is the greatest victim." Gandhi discussed the need for civil disobedience and in October 1940 wrote: "We want to tell the people of India that, if they will win Swaraj [Independence] through non-violent means they may not co-operate militarily with Britain in the prosecution of the war." (15)

On 14 July 1942, the Indian Congress approved a resolution declaring "the immediate ending of the British rule in India" which became known as the "Quit India" resolution. Gandhi was arrested by the British and incarcerated from August 1942 to May 1944. During that incarceration, Gandhi's wife, Kasturba, who was also in prison, died in February 1944. In response to a question about the trials he was facing, he maintained that his faith was strong: "I do not try to analyze God, I go behind the relative to the Absolute and I get my peace of mind…man becomes man by becoming a tabernacle of the Divine." (16)

For Gandhi, Indian Independence meant adopting Ramarajya, (i.e., "the Kingdom of God on Earth") by economic means as well as political and moral. For him, economic development meant "entire freedom from British capitalist and capital, as also their Indian counterpart. This means capitalists share their skill and capital with the lowliest and least." By moral he meant "from [the use of] armed defense forces…mankind must recoil from the horrors of war."(17)

On 15 August 1947, India became independent. Gandhi failed to establish Hindu and Muslim unity and did not succeed in integrating the Harijan (Untouchables) into the Hindu community in the way he wished. Dr. Bhimrao Ambedkar (1891–1956) preferred the name Dalits (the oppressed) and encouraged a mass conversion to Buddhism. Dr. Ambedkar, who was born a Dalit, formally adopted Buddhism toward the end of his life. He had been educated in the USA and England and became India's first Law and Justice Minister and the architect of the Constitution of India. He was critical of Gandhi's efforts to improve the condition of the Untouchable caste since Dr. Ambedkar believed that only the "annihilation of castes" would resolve the inequalities in Indian society that had subjected the Dalits to poverty and misery.

With the unleashing of the destructive forces of partition and the death of one million Hindus and Moslems, Gandhi asked: "What sin must I have committed that He should keep me alive to witness all these horrors." Facing his last fast, he wrote: "It is my belief that death is a friend to whom we should be grateful." He added: "so I am not yet a Mahatma." On 30 January 1948, Mohandas Gandhi was assassinated by a Brahmin, a nationalist Hindu who was convinced that India needed to be militarily strong and that Gandhi did not represent the kind of strength that was needed to create a strong and virile Indian nation.

Despite his own self-recrimination, Gandhi became popularly known as the Father of the Nation, and his life and writings inspired civil rights leaders, including Martin Luther King and Cesar Chavez, as well as Nelson Mandela in South Africa, who were intent on creating more just societies. Spiritual writers, such as Thomas Merton were also inspired, as were countless others who continue to take up the mantle of Nonviolence in rallies and demonstrations.

Conclusion:

Although Ireland and India were thousands of miles apart, both countries were driven by a compelling need to gain independence from Britain. Leaders, such as Gandhi in India and Terence MacSwiney in the Republic of Ireland, were willing to fast unto death. MacSwiney died in prison. In Northern Ireland, Bobby Sands, with nine others, was willing to fast to death in order to gain a united Ireland, although their intentions, prior to prison, had included violence. But Gandhi was deeply convinced that means and ends had to be consistent and resorted to fasting as an ascetic discipline for the purpose of gaining Indian independence. But for the 10 young men in the Maze prison in Northern Ireland, fasting resulted in their death. Yet, their sacrifice became a spur for other leaders, including John Hume (1937–2020) who insisted on nonviolent practices and policies in his organizing efforts. This insistence contributed to the attainment of the Good Friday Agreement of 1998, and, for his efforts, John Hume co-shared the Nobel Peace Prize and received the Gandhi Peace Prize and the Martin Luther King Award.

References

[1] Strecker, Georg: The Sermon on the Mount, an Exegetical Commentary, Translated by O.C. Dean, Abingdon Press, Nashville, 1988, (p28–30)

[2] https://www.catholicireland.net/saintoftheday/st-adomnan-eunan-624 -704-9th-abbot-of-iona/

[3] Dillon, Myles: "Celts & Aryans" 1975, Indian Institute of Advanced Studies, Simla, India. Pp 130- 13

[4] https://en.wikipedia.org/wiki/Terence_MacSwiney

[5] https://en.wikipedia.org/wiki/1981_Irish_hunger_strike

[6] https://www.projectmadurai.org/pm_etexts/pdf/pm0153.pdf

[7] Tolstoy, Leo: The Kingdom of God is Within You, Wildside Press, 2006,

[8] Tolstoy, Leo: Writings on Civil Disobedience and Nonviolence, New Society Publishers, 1987

[9] Frydman, Maurice, interview from "MK Gandhi, The Last Eighteen Years", Sterling & Ruth Olmstead & Mike Heller, (p 94–96)

[10] The Last Eighteen Years, (p 115)

[11] The Last Eighteen Years (p 152–155)

[12] The Last Eighteen Years (p 165)

[13] The Last Eighteen Years (p 169)

[14] Barash, David, Approaches to Peace, p. 229, Oxford University Press, Fourth Edition, 2018 from The Mind of Mahatma Gandhi, edited by R.K Prabhu & U.R Rao.

[15] Barash, David, Approaches to Peace, p. 231

[16] Bhagavad-Gita, translated by Swami Prabhavananda & Christopher Isherwood, Intro by Aldous Huxley, Signet Classic, July 2002 (p 29)

[17] https://en.wikipedia.org/wiki/Deep_ecology

14

Careers in Tolerance

Anjum Malik

Alhambra US Chamber

Tolerance is an urgent need at the local and global scale, but tolerance does not simply happen on its own. Instead, it benefits from formally educated professionals devoted to working for its advancement. The great demand for education to support and empower such professionals has been evident for many years. Almost a quarter of a century ago, Dr. Betty Reardon, Founder and Director of the Peace Education Center and Peace Education Graduate Degree Program at Columbia University, wrote the following in a 1997 publication:

"Any culture is fundamentally the result of learning. Education is that learning which is planned and guided by cultural values. A culture of peace thus requires an education planned and guided by the values of peace, human rights, democracy and, at its very core, tolerance."[1]

At the *International Conference on Tolerance* in Malta, October 2019, I had the honor of curating and moderating a panel on Careers in Tolerance that explored a range of professional options for those interested in promoting tolerance professionally. We examined many of the dynamics underpinning those careers. The panel format is, of necessity, broad rather than deep. This chapter allows for the luxury of presenting some of the topics the panelists and I would like to have explored in greater detail.

- Concepts of tolerance: conceptual career issues in bettering the world
- Tolerance and the arts
- Tolerance and the market: careers in international diversity

---

[1] Betty A Reardon, "Tolerance: the Threshold of Peace; a Teaching/Learning Guide for Education for Peace, Human Rights and Democracy" (UNESCO, 1994).

14.1 Concepts of Tolerance: Conceptual Career Issues in Bettering the World

Those undertaking formal study in the field of tolerance will have a range of career opportunities upon graduation. With an education centered on tolerance and peace issues, these options include working as a diplomat or field officer. Opportunities with intergovernmental agencies or in the non-profit sector include director, communication specialist, policy officer, legal advocate, project manager, conflict resolution expert, trauma specialist, human resources, professor, or researcher. Those options, it should be noted, include only traditional paths. A number of other possibilities will be explored in other sections.

14.1.1 Geographical Possibilities: Mapping Out the Future

The demand for careers in tolerance is global in scope. Developing countries gain benefit from highly skilled, mission-driven professionals with practical knowledge of the problems confronting these nations. As author Cecilia Milesi explained in a 2019 article titled, "The Role of South-South Cooperation in Realizing the Vision of Peace and Development for All," decision-making power, knowledge, and resources lead to the least amount of division, conflict, and lack of understanding when taken from a horizontal approach.[2] When policy-makers are grounded in a conceptual and practical understanding of how policies affect communities, tolerance is well served. Voices of logic, pragmatism, and compassion are necessary to address complex issues with multiple stakeholders.

Developed countries, too, benefit from a cadre of tolerance-educated professionals as they seek to expand their efforts globally. For reasons beyond the scope of this chapter, developed nations are prone to reductionist or oversimplified perceptions of problems elsewhere. Having people educated and experienced in tolerance issues safeguards against this tendency and increases the likelihood that the policies of developed nations are beneficial.

14.1.2 Skills and Contributions from an Education in Tolerance

On average, those with formal education are more aware of complexity and more capable of bringing people together to make constructive, enduring

---

[2]Cecilia Milesi, "The Role of South – South Cooperation in Realizing the Vision of Peace and Development for All," UNSSC, September 12, 2019.

change to the problems confronting the world. Those who pursue an education in tolerance studies, in particular, gain strategic competencies sharpened by experience, such as teamwork, conflict resolution, interpersonal relations, civic engagement, global citizenship, and leadership within a variety of areas including economic development, environmental studies, human security, migration studies, political philosophy, social policy, and urban studies.

Those with bachelor's degrees in tolerance-related fields are well served by continuing their education in graduate programs. The resulting knowledge and skill sets, complemented by personal passion, position them for thriving careers, as well as enhanced opportunities to promote tolerance.

14.1.3 Noble Leaders: Best Practice for Tolerance

The world still honors the late South African President and Nobel Peace Prize laureate Nelson Mandela for his accomplishments in conflict resolution, race relations, gender equality, and human rights. His work stands as a powerful legacy and inspirational example for what a life lived promoting tolerance could look like. The same is true of leaders such as the Dalai Lama, Martin Luther King Jr., and Mahatma Gandhi. While not traditionally referred to in these terms, their lives are an example of what in other fields would be called "best practice."

As in any other discipline, it is important for those seeking careers in tolerance to study best practice. The biographies and accomplishments of these leaders, and others like them, serve as grounds for education and advancement in conflict resolution and tolerance building, as well as the practical issues that must be confronted to ensure that such advancements are enduring. From the theoretical to the applied, their historical lessons strengthen current ways of working with others.

14.1.4 Professional Skills at the Next Level

Advanced degrees provide specialization within broader fields. A master's in Global Tolerance creates new opportunities for advancing academic and professional skill sets. Demand for such rigorously trained individuals will remain high. There are many gaps in tolerance and human rights around the world that need to be addressed.

14.2 Tolerance and the Arts

What is the role of the artist in promoting tolerance? A convenient place to begin answering that question is Classical Greece. While Thucydides may not have mentioned it in his *The History of the Peloponnesian War*, modern scholars see identity and culture as prime drivers of conflict. Often, what may masquerade as clashes over politics or trade actually stem from those more primal divisions,[3] in this case, between Ionian Athenians and the Doric Spartans. Historically, war and conflict have often been byproducts of community building.

Two and a half millennia later, culturally driven conflict is alive and well in our postmodern world. The Yemeni Civil War, the breakup of Yugoslavia, the Rwandan genocide, and the Israeli–Palestinian conflict are just a few of the highest profile examples. *It becomes increasingly clear that peace depends on tolerance*, rather than the two existing as cofactors. Peace is sustainable only through the embrace not only of multicultural attitudes but also of multicultural practices and sensitivity to the needs of others.

Cultural globalization, through the diffusing of customs, meanings and ideas, has enormous potential to create an environment receptive to multiculturalism and, therefore, tolerance and, ultimately, peace. For at least the past two centuries, one of the main methods in which cultures communicate with each other is through the arts.

This brings us full-circle to the role that jobs within the arts play in fostering tolerance. In several ways, artists are in a unique and enviable position in this regard. First, unlike non-governmental organizations (NGOs), intergovernmental organizations (IGOs), or international business, artists generate output that people from diverse backgrounds actively and eagerly seek out. Second, while opportunities such as MFA programs exist and can certainly be beneficial, for those with the appropriate talents, no formal schooling is necessary to become an artist promoting tolerance.

One caveat is necessary. Many societies tend to assume the artist is a friend of tolerance and, certainly, there are countless examples of such. However, there is nothing that inherently makes it so. Indeed, there are instances of art being used to drive division. In addition to the hyperbolic examples of the 1930s propaganda, there are subtler occurrences, even among beloved classics, such as the St. Crispin's Day Speech in Shakespeare's *Henry V* or the Battle of Lake Peipus scene in Sergei Eisenstein's film *Alexander Nevsky*. The use of art as a vehicle for tolerance requires a conscious choice by the artist.

---

[3] John Alty, "Dorians and Ionians," The Journal of Hellenic Studies 102 (1982): p. 1.

This section considers the media of acting, music, and literature. That should not be taken as an attempt to minimize the opportunities for careers in tolerance offered by other media.

14.2.1 Film and Television

Film and television have the capacity to directly reach vast numbers of people. While stage lacks the broad access of electronic media, most cultures assign drama a respectability and gravitas absent from their electronic counterparts, meaning its impact can be stronger upon those it does reach. From the 1920s until the present day, Hollywood has exposed billions of people around the world to American norms and values. More recently, Bollywood and filmmaking centers such as Hong Kong and Nigeria have also been vehicles for sharing culture with a global audience.

Film and television are especially promising choices for those seeking careers in tolerance. Because of the massive technical, commercial, and administrative support structures required by these media, they employ more people than most other creative industries and are full of opportunities for those interested in the arts as vehicles for tolerance but whose talents run in a different direction than performance.

14.2.2 Music

Music is unique among artistic media, in that at least part of its communication is free from dependence on language. While lyrics do matter, instrumentation and vocals can still be appreciated even in the absence of a shared language between performer and listener. And its effectiveness in communicating cultural information is unquestioned. Nearly 70 years after the appearance of Elvis and 60 years after The Beatles, rock 'n' roll has become part of the global iconography of freedom, youth, and rebellion. The high-energy and social awareness of hip hop has been adopted by cultures around the world as a vehicle for commenting on local concerns and issues.

America has no monopoly on the power of music. In recent decades, J-Pop and K-Pop have become vehicles for sharing information about and appreciation of the cultures of Japan and Korea with global audiences who otherwise might never have accessed such information. The genre known as *world music* is particularly noteworthy in this regard. While in some cases it is a straightforward vehicle for transmitting information about a single culture, in others, it becomes a pallet allowing musicians to create music weaving together elements from multiple cultures, explicitly rather than implicitly making a statement about tolerance and multiculturalism. As expressed by Deborah Pacini Hernandez, a scholar of the socio-cultural impact of music, world music (1) blurs the lines of cultural comparison and (2) develops effective alliances among different cultures.[4]

Music also has a practical appeal to artists interested in promoting tolerance. Unlike film, television, or literature, in the age of the Internet, music allows small groups of people or even a lone individual to make an impact. Recording and mixing their own music and making it available via low-cost, low-barrier platforms, musicians have the potential to reach a global audience with minimal resources.

14.2.3 Literature

Centuries before film, television, or recorded music, cultures were already sharing ideas and learning about each other through literature. More recently, the written word has served as an active tool for cultural transmission rather than a passive vehicle. Foreign language instructors have embraced literature's potential as a teaching tool and an incentive for students to absorb their lessons. Literature has not only been found to promote communicative competence but also exposure to culture.[5]

In terms of accessibility and career opportunities, literature exists somewhere between acting and music. While self-publishing tools are ubiquitous, they still have not found the same acceptance or success as their musical equivalents. And while publishing houses, marketing firms, and translators offer opportunities for those whose talents do not run to writing, they are not so massive as the support structures for television and film.

14.3 Tolerance and the Market: Careers in International Business and Development

International business may not be what most people think of when they hear the words "careers in tolerance." Nevertheless, not only have

---

[4]Deborah Pacini Hernandez, "A View from the South: Spanish Caribbean Perspectives on World Beat," The World of Music 35, no. 2 (1993): p. 48.

[5]Daniel Shanahan, "Articulating the Relationship between Language, Literature, and Culture: Toward a New Agenda for Foreign Language Teaching and Research," The Modern Language Journal 81, no. 2 (1997): p. 164.

international business and tolerance become intertwined, *international business profitability and tolerance are inexorably connected.*

Two factors have put international business at the vanguard of promoting tolerance, making such careers an appealing if less obvious road for those interested in tolerance. First, the number of people employed in international business is huge compared with those employed in IGOs or NGOs. This means, for most people, international business is their primary point of exposure to people from other countries and cultures. Second, significant overlap exists between factors that lead to international business success and those that promote tolerance. The most notable of these include diversity, teamwork, emotional intelligence, and women's empowerment.

14.3.1 Cultural Diversity

Diversity, a condition of systematized tolerance, has become the norm for international business. By pushing for tolerance not only within the business but within the broader community, business people can promote tolerance along multiple vectors. Practicing cultural diversity is a key aspect of business development in the 21st century. It has become vividly important for businesses as a competitive advantage in expanding reach and developing marketable products. Trends of increasing population mobility and migration have led to constant interactions with people of different cultures.[6] Cultural diversity positively affects the host region, providing different skills and services and a positive impact on regional growth and income.[7] Additionally, innovations are more likely to occur in a culturally diverse environment.[8]

14.3.2 Teamwork

When team members come from diverse backgrounds, as is the norm in international business, teamwork should become an exercise in applied tolerance. Whatever the backgrounds of its members, teams have a common goal and a shared desire for success. This requires respect for diverse communication styles as well as an ability to allow each member to contribute to their fullest extent in pursuit of the overall goal.

---

[6]Gianmarco IP Ottaviano, and Giovanni Peri. "The economic value of cultural diversity: evidence from US cities." Journal of Economic geography 6, no. 1 (2006): 9–44.

[7]Annekatrin Niebuhr. "Migration and Innovation: Does Cultural Diversity Matter for Regional R&D Activity?" Papers in Regional Science 89, no. 3 (2010): 563–585.

[8]Richard Florida, "The Economic Geography of Talent," Annals of the Association of American Geographers 92, no. 4 (2002): pp. 743–755.

14.3.3 Emotional Intelligence

Emotional intelligence is the capacity to understand and work with emotions, reflecting both an individual's own emotional state and an awareness of the emotions around them. Emotional intelligence skills are key to working cross culturally, and interdisciplinary with different professions and sectors. It enhances skills such as self-awareness, self-regulation, motivation, empathy, and social skills. All of these skills have applications in the business environment. While baseline emotional intelligence varies between individuals, it is a teachable and practicable capacity.

14.3.4 Women's Empowerment

Women's empowerment is a crucial factor in modern international business. A company that does not fully empower half of its workforce is at a significant competitive disadvantage to those that do. Research shows that companies with women in leadership positions and on boards of directors are more profitable than those without. This practice is especially embraced by newer companies and those with younger workforces, such as startups. However, it has made inroads everywhere and seems likely to continue to do so, as those enterprises who cannot adapt get left behind.

14.3.5 Careers in Tolerance-Related Fields

While we hope readers appreciate the deeper explorations of topics presented in this chapter, we understand that they may also be interested in the more practical aspects of our panel, including the tolerance-driven careers which are available. That list is provided below:

- Human Rights Officer, United Nations Office of the High Commissioner for Human Rights
- Project Officer, Catholic Relief Services
- Neighborhood Action Planner, City Commission
- Researcher and International Relations Expert, Institute for Trade Studies and Research
- Chief of Media and Public Information, United Nations Development Program
- Disarmament, Demobilization, and Reintegration Specialist
- Founder and Director, Wi'am Center (Palestinian Conflict Resolution Center)
- Conflict Transformation Adviser, Danida Human Rights Program

- National Coordination Officer for Peacebuilding, United Nations Mission
- Director, Complaints and Legal Services, Ugandan Human Rights Commission
- Senior Analyst, Office of the Prosecutor, International Criminal Court (ICC)
- Legal Officer, United Nations International Criminal Tribunal
- Program Manager — Iraq Refugees, Swiss Agency for Development
- Psychologist, Alliance for Community Care, specializing in gay youth and refugees
- Coordinator, Restorative Justice Program, Conflict Mediation Services
- Victim Contact Worker and Mediator, Restorative Justice Program
- Refugee Resettlement Program staff member, Catholic Charities
- Trust Fund Manager, U.N. Trust Fund for the Elimination of Violence Against Women
- Program Officer for Africa, National Endowment for Democracy
- Researcher, Conflict and Peace Analysis Unit, Centre for Policy Alternatives[9]
- Communication Coordinator/Assistant
- Community Development Organizer
- Congressional or Legislative Aide Educator
- Human/Civil Rights Officer
- Journalist
- Media Critic/Specialist
- Program Assistant/Associate/Coordinator
- Public Advocate/Activist
- Public Relations Officer
- Research Assistant
- Speechwriter
- Attorney
- Community Mediator
- Consultant (Independent)
- Editorial Columnist/Lecturer
- Executive Director of Non-Profit or Non-Governmental Organization
- Director of Non-Profit Agency or Institute
- Governmental Advisor
- Human Resources/Personnel Manager

---

[9]"Careers: Peace and Conflict Studies" (UMass Lowell), accessed July 5, 2020.

- International Negotiator
- Legal Advocate (poor, refugees, women, civil rights)
- Labor Negotiator
- Minister
- Ombudsperson
- Organizational Trainer/Facilitator Policy Analyst
- Professor
- Program Manager
- Research Associate
- Human Rights advisor for the UN. Job postings can be found on the UN website, usually acquired very fast. An adviser is usually expected to be sent to various nations around the world to a head office and integrate human safety and protection of human rights in crisis areas, and to also support the head humanitarian coordinator in the area.
- Protest coordinator for organizations such as Human Rights Watch, Amnesty International, UNICEF, IFHR, Freedom House, etc.
- App development for organizations such as the UN, Human Rights Watch, and/or the Guardian Project
- Data Analyst
- Regional Monitor (e.g., for Human Rights) Writer/Reporter (Free-lance)[10]
- Policy Developer
- Anti-Corruption Investigator
- International Diplomat
- International Crisis Relief Coordinator/Employee
- Investigative Journalist
- Peacekeeper
- Policy Evaluator/Analyst
- Legislative Advisor
- Economic Development Adviser/Analyst
- Human Rights Lawyer
- Human Rights Liaison to companies/governments
- Diversity Advocate
- Marketing Developer/Researcher geared toward tolerance and human rights
- Public Speaker/Advocate for human rights and tolerance
- Civil Rights Advocate

---

[10]"Careers in Peace & Conflict Studies" (University of Utah), accessed 5 July 2020.

Figure 14.1 Career Opportunities Panelists shown with H. E. Dr. Ahmed Al Jarwan, President, Global Council for Tolerance and Peace. Left to Right: Tony Culley Foster, Orlando Kelm, Liesl Riddle, Hassan Diab, H. E. Ahmed Jarwan, Anjum Malik, Oddgeir Tveiten, Ida Beerhalter, and Lucy Wess.

Acknowledgements

Thanks to all our panelists for making the panel, and now the chapter, such a success. Additionally, I am grateful to Kim Weichel, and the Alhambra US Chamber team — Jonathan Black, Lola Amaya, Kaitlyn Castro, Seth Johnson, Isuru Waduge, and Abigail Weigel for their help in transforming the panel into a chapter.

References

Albrecht, Milton C. "The Relationship of Literature and Society." *American Journal of Sociology* 59, no. 5 (March 1954): 425–36.

Alty, John. "Dorians and Ionians." *The Journal of Hellenic Studies* 102 (1982): 1–14. https://doi.org/10.2307/631122.

"Careers in Peace & Conflict Studies." University of Utah. Accessed July 5, 2020. https://peace.utah.edu/about/careers.php.

"Careers: Peace and Conflict Studies." UMass Lowell. Accessed July 5, 2020. https://www.uml.edu/fahss/peace-and-conflict-studies/careers/careers.aspx.

Das, J P. "Cultural Deprivation: Euphemism and Essence." *The Journal of Educational Thought* 5, no. 2 (August 1971): 80–89.

Dooley, Michael P., David Folkerts-Landau, and Peter Garber. "The Revived Bretton Woods System." *International Journal of Finance & Economics* 9, no. 4 (2004): 307–13. https://doi.org/10.1002/ijfe.250.

"Education For Peace: Top 10 Ways Education Promotes Peace" (Central Asia Institute, December 23, 2017), https://centralasiainstitute.org/top-10-ways-education-promotes-peace/.

Elliott, David J. "Music as Culture: Toward a Multicultural Concept of Arts Education." *The Journal of Aesthetic Education* 24, no. 1 (1990): 147–66.

Fish, Jefferson M. "Tolerance, Acceptance, Understanding." Psychology Today. Sussex Publishers, February 25, 2014. https://www.psychologytoday.com/us/blog/looking-in-the-cultural-mirror/201402/tolerance-acceptance-understanding.

Florida, Richard. "The Economic Geography of Talent." *Annals of the Association of American Geographers* 92, no. 4 (2002): 743–55. https://doi.org/10.1111/1467-8306.00314.

"General Assembly Declares 18 July 'Nelson Mandela International Day.'" UN, November 10, 2009.

United Nations. https://www.un.org/press/en/2009/ga10885.doc.htm.

Guilbault, Jocelyne. "Interpreting World Music: A Challenge in Theory and Practice." *Popular Music* 16, no. 1 (January 1997): 31–44.

Hernandez, Deborah Pacini. "A View from the South: Spanish Caribbean Perspectives on World Beat." *The World of Music* 35, no. 2(1993): 48–69. https://www.jstor.org/stable/43615566.

"In India, Ban Pledges UN Commitment to Gandhi's Vision of Peace, Tolerance, Dignity for All." United Nations, January 11, 2015. United Nations. https://www.un.org/sustainabledevelopment/blog/2015/01/india-ban-pledges-un-commitment-gandhis-vision-peace-tolerance-dignity/.

Kebede, Tibebu Shito. "Theoretical Debates on the Cultural Consequences of Globalization and Its Implication in the Light of the Ethiopian Culture." *International Journal of Political Science and Development* 7, no. 9 (November 2019): 287–93. https://doi.org/10.14662/IJPSD2019.160.

Lencioni, Patrick. "Peace, Discipline and Teamwork." The Table Group, November 2014. https://www.tablegroup.com/hub/post/peace-discipline-and-teamwork/.

Long, Nicole. "How to Be Tolerant in the Workplace." Small Business. Hearst Newspapers, November 21, 2017. https://smallbusiness.chron.com/tolerant-workplace-24838.html.

Milesi, Cecilia. "The Role of South – South Cooperation in Realizing the Vision of Peace and Development for All." UNSSC, September 12, 2019.

https://www.unssc.org/news-and-insights/blog/role-south-south-cooperation-realizing-vision-peace-and-development-all/.

Mukherjee, Manjari. "7 Reasons Why Bollywood Is Better Than Hollywood." RVCJ Media, October 8, 2017. https://www.rvcj.com/7-reasons-bollywood-better-hollywood/.

Muratbekova-Touron, Maral. "From an Ethnocentric to a Geocentric Approach to IHRM." *Cross Cultural Management: An International Journal* 15, no. 4 (2008): 335–52. https://doi.org/10.1108/13527600810914139.

Onu, Ben O. "Music and Intercultural Communicaiton." Essay. In *Intercultural Communication and Public Policy*, 195–212. Port Harcourt, Nigeria: M & J Grand Orbit Communications, 2016.

Ottaviano, Gianmarco I.P., and Giovanni Peri. "The Economic Value of Cultural Diversity: Evidence from US Cities." *Journal of Economic Geography* 6, no. 1 (2005): 9–44. https://doi.org/10.1093/jeg/lbi002.

Rep. *Positive Peace Report 2019: Analysing the Factors That Sustain Peace.* Institute for Economics & Peace, October 2019. http://visionofhumanity.org/app/uploads/2019/10/PPR-2019-web.pdf.

Ralph, Natalie. "Can Corporations Support Peace?" Institute for Human Rights and Business, December 8, 2015. https://www.ihrb.org/other/can-corporations-support-peace.

Reardon, Betty A. Publication. *Tolerance: the Threshold of Peace; a Teaching/Learning Guide for Education for Peace, Human Rights and Democracy.* UNESCO, 1994. https://unesdoc.unesco.org/ark:/48223/pf0000098178.

Shanahan, Daniel. "Articulating the Relationship between Language, Literature, and Culture: Toward a New Agenda for Foreign Language Teaching and Research." *The Modern Language Journal* 81, no. 2 (1997): 164–74.

Sunder, Madhavi. "Bollywood/Hollywood." *Theoretical Inquiries in Law* 12, no. 1 (2011): 275–308. https://doi.org/10.2202/1565-3404.1269.

"TANYA ST VAL - 'DOUCINE' FEAT. VICTOR O." *Radio Culture Outre Mer*, July 4, 2016. http://radiocultureoutremer.net/videos/tanya-st-val-doucine-feat-victor-o-17.

Thucydides. *History of the Peloponnesian War*. Edited by M. I. Finley. Translated by Rex Warner. Penguin Books, 1972.

Trivedi, Harish. "From Bollywood to Hollywood: The Globalization of Hindi Cinema." Essay. In *The Postcolonial and the Global*, 200–210. Minneapolis , MN: University of Minnesota Press, 2008.

"Using Literature - An Introduction." TeachingEnglish. British Council and BBC World Service. Accessed July1, 2020. https://www.teachingenglish. org.uk/article/using-literature-introduction.

Index

About the Editor

Dr. EL Zein, Director General of Governance and Sustainability Center, at (UBT), Jeddah, Saudi Arabia. She held many positions such as Dean of Scientific Research at UBT, research scientist at KAUST, faculty member at Dar AL Hekma University and associate researcher at IEMN, France.

She is currently the vice president of the Global Council for Tolerance and Peace for scientific and academic affairs.

Her research interest is currently in Energy Conversion and Energy Storage. The main objective of Dr. El Zein research is to develop an eco-green Solar Cells with high efficiency and long durability. In addition, she is exploring new materials such as kesterite, perovskite, and protein to be used as a light absorber for Solid State Sensitized Solar Cells. She is also exploring Printed Metal Oxide Batteries. She has a textbook entitled "Nanostructured materials for Photovoltaic applications" and she has many publications in this field. Additionally, she published in many international journals and has one patent filed in the USA related to materials for Solar Cells.

Dr. El Zein is a grand Judge reviewer and examiner in many international associations and a renowned guest speaker at many international conferences on renewable energy and nanotechnology. She is a reviewer for many international, peer-reviewed journals, the chair or co-chair and on committees of different international conferences. Dr. El Zein is a senior member of IEEE, member of ACS, MRS, SPIE, ECS, IET, AUTM, AAAS, AASBC, ECS and Lebanese Engineering Syndicate.

Dr. Basma graduated from the University of Lille, France with a Ph.D. in Nanotechnology Engineering with High Distinction for her research Zinc Oxide Nanostructures for Photovoltaic Applications. Her master's degree was from the Lebanese University, Lebanon in the field of Electrical and Electronics Engineering with Distinction.

Dr. El Zein was acknowledged for her remarkable contribution to the field of research, teaching, community service, and industry in Saudi Arabia, and she was granted 2 funded research for her research on Nanomaterials for 3rd generation Solar Cells in 2013 and 2014.

Raised on the teachings of peaceful co-existence, **Ahmed Bin Mohammed Al Jarwan** is the president and founder of the Global Council for Tolerance and Peace in Malta since 2017 and president of the Arab Experts Union since 2018. Earlier, he was president of the Arab Parliament from 2012 to 2016, member of the Federal National Council of the UAE from 2011 to 2018, member of the board of directors of the Sharjah Chamber of Commerce and Industry from 2008 to 2012, and founder and manager of Sharjah Old Cars Club and Museum from 2007 to 2012.

Mr. Al Jarwan played a significant role in tolerance, peace and relationship with the United Nations, UNESCO and the Arab League. He was a key player and essential participant in all activities of international organizations and in national and international parliamentarians' meetings, receiving great recognition from dignitaries worldwide. He earned the 2020 Mahatma Gandhi International Award and has been recognized by the leaders of such countries as Russia, Colombia, Argentina, Albania, Kosovo, Djibouti, Comoros and Lebanon for his achievements regarding human rights, tolerance, peace, sustainable development, and women and youth rights.

Looking forward, Mr. Al Jarwan is raising the value of tolerance and spreading the culture of peace, fighting discrimination, violence, terrorism and racism, and strengthening principles of tolerance to achieving peace. He intends to continue advocating for humanitarian issues and being a great Emirati role model, reflecting the successful Emirati culture that supports tolerance and peace.

About the Authors

Timothy Reagan is Professor of Applied Linguistics and Foreign Language Education in the College of Education and Human Development at the University of Maine. He has held senior faculty and administrative positions at the University of Connecticut, the University of the Witwatersrand, Central Connecticut State University, Roger Williams University, Gallaudet University, and Nazarbayev University in Astana, Kazakhstan. His primary areas of research are applied and educational linguistics, language policy and planning, and issues related to sign languages. Reagan is the author of more than a dozen books, including *Linguistic legitimacy and social justice* (2019, Palgrave Macmillan), *Language planning and language policy for sign languages* (2010, Gallaudet University Press), *Language matters: Reflections on educational linguistics* (2009, Information Age Publishing), *Critical questions, critical perspectives: Language and the second language educator* (2005, Information Age Publishing), and with Terry A. Osborn, *World language education as critical pedagogy: The promise of social justice* (forthcoming, Routledge) and *The foreign language educator in society: Toward a critical pedagogy* (2002, Lawrence Erlbaum Associates). He is also the author of more than 150 journal articles and book chapters, and his work has appeared in *Arts and Humanities in Higher Education, Critical Inquiry in Language Studies, Educational Foundations, Educational Policy,*

Educational Theory, Foreign Language Annals, Harvard Educational Review, Language Policy, Language Problems and Language Planning, Multicultural Education, Sign Language Studies, and *Semiotica.* He served as co-editor, with Terry A. Osborn, of *Critical Inquiry in Language Studies* from 2004–2007, and as Editor-in-Chief of *Language Problems and Language Planning* from 2014–2018.

Karina V. Korostelina is a Professor and a Director of the Program on History, Memory and Conflict and a co-director of the Program on Preventing Mass Atrocities at the Carter School for Peace and Conflict Resolution, GMU. Her recent interests include the study of national and community resilience, reconciliation, and role of memory and history in conflict and post-conflict societies. She has been Fulbright New Century Scholar and fellow at the Woodrow Wilson Center, the Eckert Institute, National University of Singapore, East-West Center, Institute for Advanced Studies at Waseda University, Northeast Asia Foundation, Central European University, and the Bellagio Center of the Rockefeller foundation. She has received over 40 grants from such Foundations as MacArthur, Luce, Spencer, Ebert, and Soros, the US Institute of Peace, US National Academy of Education, Bureau of Educational and Cultural Affairs of USDS, USAID, INTAS, IREX, and Council of Europe. The results of her research are presented in more than 90 articles and chapters. She is an author or editor of 16 books including authorship of *Trump Effect* (2016), *International Insult: How Offence Contributes to Conflict* (2014), *Constructing Narrative of Identity and Power* (2013), *History Education in the Formation of Social Identity* (2013), *Why they die?* (2012), *The Social Identity and Conflict (2007).* Among her edited books are: *History Can Bite - History Education in Divided and Post-War Societies* (2016), *History Education and Post-Conflict Reconciliation* (2013), *Forming a Culture of Peace* (2012), *Civilians and Modern War (2012), Identity, Morality and Threat (2006).*

Professor Stephen Dobson was born in Zambia, grew up in England and moved to Norway in the 1980s.

He is currently Dean of the Wellington Faculty of Education, Victoria University of Wellington (New Zealand), a Guest Professor in Lifelong Learning at Inland Norway University of Applied Science and an Adjunct Professor at the University of South Australia.

Dobson possesses deep knowledge and experience of education eco-systems across SE Asia, Europe and Scandinavia.

He holds a PhD in refugee work, a second PhD in assessment, and a Magistergrad in sociology complements his later studies. By qualification and practice a Professor of Assessment, he has held many workshops for in-service teachers in ECE, primary, secondary and tertiary levels of education and Lifelong Learning.

Dobson has published extensively on student assessment, youth studies, ethnicity, research methods and educational philosophy.

Sofia Herrero Rico holds a PhD with international mention by the Universitat Jaume I (UJI), Castellón Spain. Her Doctoral Dissertation *Educación*

para la Paz. El enfoque REM (Reconstructivo-Empoderador) obtained the extraordinary PhD Award in the 2012–13 academic year. She holds an international Master Degree in Peace, Conflict and Development Studies and a BA in Pedagogy.

She is the Coordinator of the UNESCO Chair of Philosophy for Peace, researcher at the Interuniversity Institute for Social Development and Peace (IUDESP) and member of the research group 030 Social Development and Peace (Philosophy, Communication, Education and Citizenship).

She has carried out research stays in the US, Portugal, Italy, Mexico and Costa Rica, and has given courses, seminars, lectures and workshops both nationally, and in Peru, Algeria, Italy, Mexico, Finland, USA and Malta.

She develops her teaching career at the UJI, where she also participates in several research and educational innovation projects related to education and cross-cutting issues such as human values, peace, recognition of each other and creativity.

She has different publications, books, book chapters and articles in specialized journals.

Gaetano Dammacco is full Professor of Eecclesiastical Law (i.e. juridical discipline of the right of religious freedom) and Canon Law. He is Director of the editorial scientific series (entitled "Diritti-Società-Religioni"), Director of scientific law and economic review (entitled "Euro-Balkan Law and Economics Review"). He is Expert in sociology of religion, Director of University Masters. Member of scientific committees of law reviews. Evaluator and member of the evaluation team of the Italian Ministry of University and Research. Visiting professor in various foreign universities (Albania, Poland, Spain). He organizes national and international conferences, especially on issues related to the protection of fundamental human rights and religious freedom. He was director of Law Department, member of the university's evaluation teams, he was responsible for international inter-university agreements. Now he is university master teacher. He is responsible for scientific research in the field of human rights and religious freedom. He is author of numerous essays and monographs (over 130) relating to numerous topics about religious freedom, human rights, the relationship between religion and economy, environmental protection, internal governance of the Catholic Church, marriage, protection of minors, religion and protection of sensitive personal data, cultural heritage, relations between states and religions, conscientious objection, dialogue between religions. One of his monographs entitled "Diritti e religioni nel crocevia mediterraneo" (Rights and Religions in the Mediterranean Crossroads) is translated into Arabic, published in Syria / Lebanon by the *Atlas* publishing house.

Mrs. Eda Çela was born on 25.01.1991 in Elbasan, Albania. She is graduated in Bachelor Program in 2012 in the Faculty of Law, University of Tirana, Albania and in the same Faculty graduated in Master of Science in Public Law in 2014. She enrolled in the National Bar School after finishing the university studies and in 2015 arranged to become a licensed lawyer. From 2017 she is a PhD student in University Aldo Moro, Bari, Italy, in the Constitutional Comparative Law

Mrs. Eda Çela has been working as a lawyer in legal offices in Tirana, Albania and also has been contacted as a legal expert in projects with OSCE and Erasmus+ Capacity Building Projects. From 2016 she is enrolled as an assistant professor for Constitutional Law, at Faculty of Law, University of Tirana, University of Elbasan "Aleksandër Xhuvani" and LUARASI University.

From January 2017 she is the Head of Coordination and International Relations Office at the University of Elbasan "Aleksandër Xhuvani". After graduation she has participated in national and international conferences and has undertaken training and different qualifications.

Roberta Santoro, associate professor of Ecclesiastical and Canonal law - Department of Political Sciences - at the University of Bari "Aldo Moro". She teaches "Systems of Relations between State and Churches", "Rights and Religions in European Societies", "Rights of Religions and Multicultural Citizenship", "Human rights and Geopolitics of Religions". She is Visiting Professor at the University of Elbasan (Albania). Topics of research mainly focus on protection of religious freedom and on relationship between law, multicultural society and confessional pluralism, relations between State and Churches, religious confessions. She is author of various publications including *"Confessional Membership and Citizenship Rights in the E.U."*; *"Conscientious Objection and Confessional Affiliation"*; *"Religious Phenomenon and Dynamics of Multiculturalism"*. She is Lecturer in various Master courses including EMCT *"Euromeditteranean Master in Culture and Tourism"* promoted by Emuni- Slovenia and Health Law. She is responsible for the Law and Civil Legislation section of the National "A-Class" Journal, and she is Co-Director of the International Scientific Journal.

Fernando Mimoso Negrão, born in Angola in November 1955.

Currently Married and father of two adult sons, studied in Lisbon and had the graduation in law in the Lisbon Classic University.

After 3 years studying to be a judge, performed the task during 20 years in several places in Portugal.

Was National Director of the Judiciary Police and National Coordinator of the drug combat before being elected Member of the Parliament.

Throughout the political exercise, was Social security, family and children's Minister, and Justice Minister.

Nowadays, is still a Member of the Parliament and Vice President of the Portuguese Parliament.

Cláudia Vaz Assistant Professor at Institute of Social and Political Sciences (Lisbon University), Researcher at the Centre for Public Administration and Public Policies (CAPP) Member of CPLP Thematic Commission on Education, Higher Education, Science and Technology (representation ISCSP)

Coordinator of *The Starting Point – Educational Experiences IN* (key words: development of human potential, inclusion, soft skills, transformation, global, non-formal) Anthropologist with published works on culture and identity of children and youth of African origin and coordination and participation in research on participatory methodologies and gender.

João Ferreira Trainee of *The Starting Point – Educational Experiences IN* Last year International Relations Student at Institute of Social and Political Sciences (Lisbon University).

Ana Carolina Reis Trainee of *The Starting Point – Educational Experiences IN* Last year International Relations Student at Institute of Social and Political

Sciences (Lisbon University) Member of the Magna Tuna Apocaliispiana (ISCSP Academic Tuna).

Hugh Curran was born in Donegal, Ireland into a Gaelic speaking family and after moving to Canada did undergraduate studies in Nova Scotia. He subsequently lived for five years as a Zen monastic and did a three-month pilgrimage to India and Japan before moving to Maine. He became a founding member of the *Morgan Bay Zendo* where he is on the Board of Directors & guides retreats and also founded the *Friends of Morgan Bay* which oversees several nature preserves. During the 1990s Hugh became the Director of a homeless shelter in Downeast Maine and has published articles on home-lessness. Since 2002 he has been a lecturer in the **Peace & Reconciliation Studies Program**. Hugh has co-written a book on local Maine history with Esther Wood and has published poetry in various poetry journals as well as compiling a classroom text*: Excerpts from Classical & Modern Writers on War & Peace. In July, 2017* he was invited to present a paper on a *"Buddhist Interpretation on the Ethics of Animal Suffering"* at St. Stephen's College, Oxford University, UK. In October, 2019 he was invited to give a talk on *Tolerance & Nonviolence* at *the Global Council on Tolerance & Peace* in Malta.

Ferdinand Gjana is Professor of International Relations and Security. He holds a PhD in International Relations and Diplomacy at Center for Strategic and Diplomatic Studies (CEDS) in Paris, France. Dr. Gjana has participated the Study of the U.S. Institute program on Religious Freedom and Pluralism in U.S. at Temple University. He has been part of many programs as visiting scholar, guest professor, keynote speaker or external reviewer at University of Northern Iowa, Harvard University, Zhejiang Normal University, Albanian Armed Forces Academy, Qatar University, United Nations, Council of Europe and European Commission. Dr. Gjana is a senior analyst for Strategic Planning, Geo-Political and Economic analysis, forecasting and risk management solutions at Wikistrat, Washington DC.

Since 2011, Dr. Gjana holds the position of the Rector at University College Bedër in Tirana, Albania.

Dr. David L. Everett currently serves as the Associate Vice President for Inclusive Excellence at Hamline University. Dr. Everett studies approaches to equity, inclusion, and diversity in the context of organizational culture, leadership dynamics, as well as interpersonal communication. His work seeks to add to current discourse by focusing on necessary individual and institutional aspects for systemic change.

As an area specialist with over 15 years of experience, Dr. Everett's approach engages and explores the impact of systemic and structural dynamics to achieve better overall results from equity, inclusion, and diversity efforts.

As a Qualified Administrator for the Intercultural Development Inventory and Certified Federal Mediator, Dr. Everett has contributed to the work of state agencies, county municipalities, and educational entities.

Anjum Malik is the Managing Partner of the Alhambra US Chamber, Executive Director Global Impact Initiative, VP HOFT Institute USA, Telemachus Mentor Global Thinkers Forum UK, Honorary Ambassador Polish Network of Women Entrepreneurs and an Associate of the Future Learning Lab Norway.

Anjum is a global education entrepreneur, consultant, mentor and presenter.

She has built an extensive network of international contacts which she leverages on behalf of her clients, connecting people and organizations for success. Along with her expertise in strategic planning, market development, negotiation and management, her work in the U.S. and multiple international agencies documents an impressive record of success.

Anjum's belief in the power of education to improve society is evident in all aspects of her professional life. She has enriched the academic, linguistic and professional development of international students, scholars and administrators. The organizations she has created and managed have enhanced the global competence of more than 170,000 students and hundreds of professionals. Within her most nurtured passion - international education - Anjum constantly seeks innovative tools to improve teaching, learning, and access. In all her endeavors, she strives to advance the empowerment of women and their increased participation in the workforce.

Anjum consults and trains on a wide variety of issues related to international education and has served as an advisor to The Initiative to Educate Afghan Women; was a founding member of The University of Texas at Austin's Global Initiative on Education and Leadership (UTGI) and the Coalition for Airline Passenger Rights; has served on various boards including the Boys & Girls Clubs of America, St. David /South Austin Hospitals (Trustee), Action for Afghan Women, and the Tex US Business Coalition; has worked with multiple educational and cultural outreach programs; and has been interviewed and quoted in national and international media.

Because of Anjum's leadership in education delivery and development, the Chamber was invited to become a founding member of President Barack Obama's Partners for a New Beginning Program.